a love of the land

CENTER BOOKS ON AMERICAN PLACES AND THE INTERNATIONAL SCENE

George F. Thompson, series founder and director

a SELECTED *love* WRITINGS OF *of* JOHN FRASER HART *the land:*

edited with an introduction

BY JOHN C. HUDSON

THE CENTER FOR AMERICAN PLACES

AT COLUMBIA COLLEGE CHICAGO

Published 2008. First edition.
Printed in China on acid-free paper.

The Center for American Places at Columbia College Chicago
600 South Michigan Avenue
Chicago, Illinois 60605-1996, U.S.A.
www.americanplaces.org

Distributed by the University of Chicago Press
www.press.uchicago.edu

16 15 14 13 12 11 10 09 08   1 2 3 4 5

Library of Congress Cataloging-in-Publication Data

Hart, John Fraser.
A love of the land : Selected Writings of John Fraser Hart /edited with an introduction by John C. Hudson.
p. cm.
Includes bibliographical references.
ISBN 978-1-930066-75-5 (hardcover) -- ISBN 978-1-930066-76-2 (pbk.)
1. Geography. 2. Landscape. I. Hudson, John C. II. Title.

G116.H37 2008
910.92--dc22

2008004833

ISBN-10: 1-930066-75-9 (hardcover)
ISBN-13: 978-1-930066-75-5 (hardcover)
ISBN-10: 1-930066-76-7 (paperback)
ISBN-13: 978-1-930066-76-2 (paperback)

Frontispiece: A horse farm in Kentucky's Bluegrass country, 1959.
Photograph by John Fraser Hart.

CONTENTS

John Fraser Hart, courtesy of the University of Minnesota.

INTRODUCTION

I did not have to think twice before agreeing to edit this collection of writings by John Fraser Hart for the Center for American Places. A living icon of American geography, Fraser Hart became the best-known geographer of his generation, an exemplar of the highest standards of geographical scholarship, and an inspiration to countless others who appreciate the example he has set. Academic societies do not bestow honors for having served as a muse, but if they did my hunch is that Fraser Hart would be among the first to be so honored. Some scholars are known for a particular idea or for a paper they wrote, while others are known for the students they produced or for the influence they had on the departments in which they made their careers. Fraser Hart has gained recognition in all of those respects, but his greatest contribution probably has been his indefatigable presence as a participant in all things geographical over a career that has spanned six decades.

He was born April 5, 1924, in Staunton, Virginia. His early years were spent in the small college town of Hampden Sydney, Virginia, where his father taught history at Hampden Sydney College. After he graduated from public school at the age of fifteen, the Hart family moved to Atlanta, where young Fraser enrolled at Emory University. He majored in classics and received an A.B. degree from Emory in 1943. Then he joined the U.S. Navy and was on active duty in the Pacific Theater from 1943 to 1946.

When the war ended, Hart returned to Atlanta and enrolled in classes at the nearby University of Georgia in Athens. His first course work was in chemistry, but he soon met Professor Merle Prunty who encouraged his interests in geography. On Prunty's recommendation, Fraser Hart enrolled at Northwestern University for graduate work, where he earned a M.A. in 1949. Prunty then hired him back to Georgia, where Hart taught while completing his doctoral dissertation in 1950. Part of a postdoctoral year spent with Professor Richard Hartshorne at the University of Wisconsin-Madison in 1951–1952 began an enduring intellectual friendship.

From 1955 to 1967 Fraser Hart was on the faculty at Indiana University, where he rose from assistant professor to professor and built an enviable reputation as a scholar who blended economic, historical, and cultural themes in the study of land and land use. At Hart's urging, Carl O. Sauer was invited to spend a semester teaching at Indiana in the early 1960s, which provided another shaping influence. As Hart's involvement with professional geography grew, he was drawn into affairs of the Association of American Geographers (AAG) and spent a year (1966–1966) in Washington, D.C., as its executive officer. He later served as editor (1970–1975) of the *Annals of the* AAG, as a member of the AAG Council (1976–1978), then as vice-president and president of the AAG (1978–1980).

In 1967, Fraser Hart moved from Indiana University to the University of Minnesota, where he remains a member of the faculty. Although no change in scholarly direction accompanied the move to Minnesota, Hart both broadened and deepened his commitment to professional geography in the years that followed. He served three years (1974–1977) as councilor of the Canadian Association of Geographers, and six years (1970–1976) as a member of the U.S. National Committee for the International Geographical Union, and he spent twenty years (1976-1996) as an editorial advisor to the *Geographical Review*. He was a John Simon Guggenheim Memorial Foundation Fellow in 1982–1983.

Among the awards and prizes Fraser Hart received in his later career are the Award for the Teaching of Geography from the National Council for Geographic Education (1971), the first Lifetime Achievement Award from the Southeastern Division of the AAG (1987), the John Brinckerhoff Jackson Prize for his book, *The Land that Feeds Us* (Norton, 1991), the Paul P. Vouras Medal from the American Geographical Society in 2005, and a Lifetime Career Award from the AAG in 2005.

A Love of the Land brings together thirteen of Fraser Hart's articles and book chapters. The title, which echoes *The Look of the Land* (Prentice Hall, 1975), perhaps his best-known book, suggests an underlying theme for his life's work, although one that probably requires explanation at the outset. Nowadays, those who "love" the land often proclaim that love loudly and stridently. They use words such as "irreparable," "pristine," and "desecrate" when speaking of human impacts on the natural landscape, and they reserve praise only for efforts to preserve or restore what they imagine used to be. Even a brief glance at the chapter titles in this book will show that Fraser Hart is not one of those people. When it comes to landscapes his tastes are thoroughly catholic, and, although he certainly does not like all that he sees in the world around him, his desire to understand the land and how it is used always has been greater than any impulse he has had to condemn what he saw. His hardheaded, statistical approach to the analysis of urban sprawl in the United States is one example (Chapters 8 and 9 in this volume). In the 1970s, he was one of the first to recognize that popular accounts and even some official government reports were wildly overstating the extent to which urban sprawl was "gobbling up" valuable farmland on the fringes of American cities. While not doubting that sprawl existed, he demonstrated that it posed no danger to the nation's food supply, as some were suggesting. Many factors cause land to go into and out of agricultural production, and urban sprawl is a comparatively minor component of the overall dynamic.

Although obviously a member of the comfortably ensconced, well-educated, suburban class himself, Fraser Hart is not a strong advocate of that group's dominant ethos. Second-home resort developments (Chapter 10) and mobile homes (Chapter 11) typically are disliked by people who say they love the land, but those kinds of settlement do exist, and in some parts of the United States they are dominant. Fraser Hart realized that we knew little about such things and, in the 1980s, he began to study them systematically. The association between socioeconomic class and resort-area styles became apparent as he traveled around the Upper Great Lake states, stopping and getting to know the resort areas not as a tourist but as a geographer in the field. Mobile homes carry an even stronger popular disapproval than do resort cabins. His research on house trailers in the 1990s happened to coincide with the brief notoriety of Tonya Harding, the champion figure skater who was guilty of underhanded tactics in her quest for an Olympic gold medal. In one of the most entertaining academic presentations I have ever witnessed, Fraser Hart addressed a session at the Association of American Geographers meeting in 1995 on the subject of house trailers and began with several newspaper quotations about Harding. She had "grown up in a house trailer," the newspapers said, simultaneously a condemnation and an explanation of her low behavior.

House trailers, resort cottages, and suburban tract homes all exist on the land, and we are better off understanding why they are there than we would be if we simply turned our backs to such things. I never heard Fraser Hart speak those words, but I expect he would agree. He probably would add, for good measure, that, because they are there, it is our duty as geographers to go out and study them. They are not really to love, however, no matter how much time we spend around them, nor would studying them lead a young person into the field of geography in the first place. One probably could make the case that the first land a geographer studies by choice is the one which he or she loves and that, the more freely chosen it is, the more likely it represents a cherished place.

For Fraser Hart, this would be the British moorlands (Chapter 3). With pride in his Scottish ancestry and mindful of the need for a geographer, circa 1950, to have a foreign area of expertise, Hart elected to write his Northwestern University doctoral dissertation on hill sheep farming in southern Scotland. His dissertation advisor was Malcolm J. Proudfoot, an accomplished scholar who, unfortunately, was little known to later generations due to his untimely death in 1955. "Uncle Mal," as Hart fondly refers to him, was a taskmaster. Proudfoot demanded not only the dissertation chapters on physical environment, land use, and farm production that would be expected in a work of that title, but he also required two case studies of individual farms that would draw the various themes together. For Fraser Hart, case studies would become an integral part of studying land, how it is used, and how locality fits into a larger picture.

Hart later used case studies of farms in articles that he published in professional geographical journals and in the several textbooks he wrote for use in primary and secondary schools; eventually he refined the approach beyond farm layouts to focus on the critical elements of decision-making and entrepreneurship that must guide a successful farming operation. Land is not simply used: it is used by individuals, for a purpose, and their strategies in doing so can teach us a great deal about why the landscape looks like it does. His book, *The Land that Feeds Us*, from which Chapter 12 is taken, was organized around more than two dozen case studies of individual farms and showed how agriculture in the eastern half of the United States changed in the decades following World War II.

Even this was not enough to capture the massive changes in American agriculture that were evident by the 1990s. Crops and livestock, the twin components of agricultural production on the "typical family farm," were becoming increasingly separated from one another and were migrating to new areas once thought impossible for their success. Many Midwestern farmers were abandoning livestock and focusing entirely on the production of grain to be sold for cash (Chapter 13). Poultry production had become a factory-like operation by the 1960s (Chapter 7), which was followed by beef in the 1970s, pork in the 1980s, and dairy and egg production in the 1990s. *In The Changing Scale of American Agriculture* (Virginia, 2003), Hart shows how individual entrepreneurs in each of these industries took chances, expanded their businesses, and changed the course of agricultural geography in the nation. "Change in the Corn Belt" (Chapter 13) is taken from Hart's 2003 book, although it is actually the second revision of an article by the same title Hart first published in the *Geographical Review* in 1986, which he later updated and published in *The Land that Feeds Us*. Change has been both rapid and relentless.

John Fraser Hart obviously has had a front-row seat from which he viewed and wrote about these many changes in American agriculture over the past half century, but no one gave him that seat nor did anyone point his way toward it. To know very much about what is happening as it happens, and to command a full enough understanding of the trends to write about them in a scholarly fashion, demands an enormous dedication to hard work, not to mention the intestinal fortitude it requires to risk being wrong, especially when you know you are in the middle of a massive upheaval. "How does he know all this?" is a frequent comment one hears after one of his meticulous presentations at professional meetings.

The answer is neither simple nor obvious. Fraser Hart is an inveterate reader of newspapers (especially the *New York Times*, which he saves in great stacks and then clips from while engaged in his single vice—watching football on television). Newspapers are a source of new ideas, but only that, and whatever one learns from reading them has to be verified. Hunches can be right, although they are often wrong, and the only way to test them is to see for yourself. For example, I would guess that one of Fraser Hart's favorite sources of new ideas about what is happening to agriculture in the Middle West has been the annual automobile trips he has made since the early 1970s from his home in the Twin Cities to the South Dakota State Geography Convention in Brookings. Something seen on the trip out can be studied in more detail on the return. I have accompanied him on some of these trips, as well as on many others elsewhere, and thus have grown accustomed to his approach.

Unlike some geographers, who have made almost a fetish out of "field work" because of the supposed higher order of insights thereby gained compared with any other mode of investigation, Fraser Hart has never stopped there. What one sees is simply an impression, as is the anecdotal information gained from conversations with a farmer or a county agricultural agent whom one happens to meet. For farming, the ultimate test of a trend is the U.S. Census of Agriculture, the compendium of detailed agricultural data published every five years that is based on an actual census of farm operators. Even this is not the end of the process, however, and any changes revealed in the data or learned in first-hand observation should be observable in aerial photographs and mapped on the ground. Once something has been seen, tallied, and mapped, one must return, tape recorder in hand, and learn more, first-hand, from the principals involved. Only then can it form the basis for a scholarly paper.

Whether one chooses to call this science or art or even journalism, there can be no doubt that the approach is comprehensive and that anyone who chooses to spend life this way has little time left over for other pursuits. Fraser Hart wastes no time. Numerous trips made to Door County, Wisconsin, where his wife Meredith and their children Laird and Anne and their families enjoy spending time in the summer, obviously led him to study that area as a geographer would. He has published several professional papers about the place, and, as I write this, Fraser Hart is at his typewriter (he does not favor computers), completing a book about Door County for the new *My Kind of* series that the Center for American Places at Columbia College Chicago debuts in 2008. A visiting professorship at East Carolina University in 1979 eventually led to three papers on changes in tobacco production. Trips to attend annual meetings of the Southeastern Division of the AAG have kept him up-to-date on changing land use in the South.

Fraser Hart once gave me a tongue-in-cheek description of geographical field work as "aimless driving about the countryside collecting miscellaneous tidbits of useless misinformation." His own practice contradicts that description by a wide margin, but he is sensitive to criticisms of regional geography, which is his favorite brand of scholarship, on the grounds that it is not "science." It is too open-ended, too inclusive of details, and (perhaps most damning of all), too downright interesting to be the work of a professional scientist. In mid-career, in 1981, he delivered his presidential address to the Association of American Geographers, calling regional geography "The Highest Form of the Geographer's Art" (Chapter 1).

Like the rest of us, Fraser Hart did not anticipate that "science" would evolve into the monster that it has become, with massive funding awarded by scientists to one another in the sacred rite called "peer review," with research agendas driven as much by politics as by scientific curiosity, all in the name of proving or disproving that the world is going to hell in a hand basket. To me, however, Fraser Hart's work represents the ideals of scientific inquiry better than most of what I see around me in the university today. The purpose of science is explanation, which begins with a hunch and not with data collection. One cannot collect evidence until there is a proposition that can be judged against that evidence. The time-honored term for this is "sense of problem," and it is a quality evident in all of Fraser Hart's papers. A hunch becomes a serious proposition if it seems to stand the test of comparison with all available evidence; it becomes an explanation only when it can be fit logically into a larger scheme of reference.

In the late 1960s, when geography was at the height of its fascination with statistical methods, researchers were more self-conscious about this process to the point that it flavored the language of research studies. In "Field Patterns in Indiana" (Chapter 5), statistical methods are used to discriminate between field sizes and fence styles in three agricultural areas of Indiana. The approach stands in sharp contrast with "Tobacco Barns and Their Role in the Tobacco Economy of the United States" (Chapter 6), which is a free-flowing account of how the barns used to store tobacco reflect both the type of tobacco stored in them and the method used to cure the product. "Tobacco Barns," published in 1961, was coauthored with Eugene Cotton Mather (1918–1999), a long-time friend and colleague with whom Hart collaborated on more than a dozen papers. Although written in different styles, both "Field Patterns" and "Tobacco Barns" are replete with hard evidence that only careful field work could have produced.

Conjecture, evidence, hypothesis, and a willingness to consider alternative explanations are the stock in trade of the good scientist. John Fraser Hart is as comfortable with those routines as anyone in the discipline, notwithstanding his belief that regional studies are the highest form of the geographer's art. It is in the writing that regional geography becomes an art, and perhaps it is for this reason that Fraser Hart has become such a well-known advocate of clear, expository writing in geography. I have heard him say more than once, "The Good Lord probably intended me to be a copyeditor." Perhaps so, but if that is the case we are fortunate that the Lord did not make His intentions known and thereby deprive the geographic profession of one of its most effective communicators.

The end product of all research should be publication, Hart believes, or more specifically, to quote "The Highest Form of the Geographer's Art," "communication is the sole purpose of scholarship." Even more strongly, he insists, "the understanding, appreciation, and interpretation of any part of the world is completely worthless until the insights one has gained have been effectively communicated to others." In the same paper he reminds us that "the uncorrupted epigram is 'Publish your results, or they will perish with you.'" There can be no doubt that he practices what he preaches. The author of more than a dozen books and well over 100 research articles, Fraser Hart probably has spent more time writing (and rewriting) than he has spent working in the field and teaching in the classroom combined.

He begins a work day with a stack of paper half a foot high, typing first drafts triple-spaced on the backsides of already used sheets. He ruthlessly applies his editorial blue pencil to what he has written, combines fragments into sentences, reworks them into longer sentences, and then into paragraphs. A well-earned break to enjoy a few puffs on one of his ancient, encrusted pipes comes only after he has mastered what he wants to say.

I have chosen to highlight some of Fraser Hart's research strategies and work habits in this introductory essay, rather than to comb over the historical record of his life, because this is a collection of a living scholar's work, not a memorial, and if it has value for a younger generation it will be to read the work itself rather than to focus on details about the man who wrote it. Now well into his eighties, Fraser Hart remains an active scholar and teacher. Although we may never see his like again, we do look forward to more and wish him all the best as his journey continues.

The thirteen essays that form this book are grouped very roughly in chronological order with the exception of Chapter 1, his Association of American Geographers presidential address delivered in 1981, which follows this introduction. Chapter 3, on the British moorlands, was written twenty-five years earlier, but in it one can see that he was already following the advice he would later give. Chapter 2 ("Reading the Landscape"), Chapter 4 ("Land Division in Britain"), and Chapter 5 ("Field Patterns in Indiana") examine the layout of the rural landscape, although in very different institutional and historical contexts. Chapter 6, on tobacco barns, and Chapter 7, on land-use change in the Georgia Piedmont, are but two of the many papers he has contributed on the geography of the American South. Chapters 8 and 9, written twenty years apart, focus on the conversion of rural land to urban uses, while Chapters 10 and 11, on resort areas and mobile homes, represent the extension of interests beyond the realm of agriculture. Chapters 12 and 13 return to the agricultural theme but with an emphasis on the role of entrepreneurship in farming and food production. The essays have been edited for length, and a few parenthetical remarks have been inserted [in brackets] to update the text or to help the nonspecialist reader, but otherwise nearly all of the original wording has been preserved.

I wish to thank George F. Thompson, founder and director of the Center for American Places at Columbia College Chicago, for his good will and support throughout this project. Amber Lautigar, Lenore Lautigar, and others on the Center for American Places staff provided invaluable assistance. I also wish to thank two anonymous referees, for their helpful comments on the text.

gallery of photographs by John Fraser Hart

Wear Cove, Sevier County, Tennessee, August 1959. Jim Fry and his grandson in front of a patch of tobacco and next to the "Indian" patch of corn, beans, and cucurbits that grew all the food that the family needed.

Wear Cove, Sevier County, Tennessee, August 1959. Wear Cove is a lush limestone valley nestled in the wooded Great Smoky Mountains of eastern Tennessee.

Clarksdale, Mississippi, August 1959. Cotton on level alluvial land in the Delta country of western Mississippi.

Leland, Mississippi, August 1959. Sharecroppers in the Mississippi Delta country lived in forlorn "cabins in the cotton patch."

East of Luverne, Alabama, April 1962. Low earthen terraces for erosion control snake through a field that will be planted to cotton. The terraces in the pasture in the foreground reveal that it, too, was once a cotton field.

North of Champaign, Illinois, October 1982. Vast fields of corn and soybeans dominate the Grand Prairie of east-central Illinois, in the heart of the Corn Belt, which stretches 800 miles east and west by 200 miles north and south.

North of Crawfordsville, Indiana, October 1956. An old-fashioned mechanical corn picker stripped the ears of corn off the stalks, and the conveyor belt carried them to the trailing field wagon.

a

love

of

the land

FIG. 1.1. Limestone bluffs on the Green Bay side of Door County, Wisconsin, 1980.

The Highest Form

CHAPTER 1

of the Geographer's Art*

Society has allocated responsibility for the study of areas to geography; this responsibility is the justification for our existence as a scholarly discipline. Most people are inherently curious, and they want to know more about the world in which they live—about their home area, about what's over the hill, about what's beyond the horizon, about what's across the sea—and it is our duty, as geographers, to satisfy their curiosity.

Unlike most disciplines, geography cannot be organized around a unique set of phenomena, but, like history, it is essentially a unique point of view. Geographers cheerfully accept and try to integrate the apparently bewildering variety of disparate phenomena that give areas their distinctive character. Geography is a science, but it is also an art, because understanding the meaning of an area cannot be reduced to a formal process. The highest form of the geographer's art is producing good regional geography—evocative descriptions that facilitate an understanding and appreciation of places, areas, and regions.

Insecurity and Scientism

Despite the high and demanding mission of geography, and the distinction with which geographers have carried out their mission, we are a surprisingly insecure lot. One of our favorite pastimes is complaining that society refuses to give us the recognition we think we deserve. If geographers in a democratic

* "The Highest Form of the Geographer's Art," *Annals of the Association of American Geographers* 72 (1982): 1–29.

society desire a better public image, we must ask people what they want from us, instead of telling them what they ought to want—although as educators we also have a responsibility for trying to raise expectations that we consider too low. Then we must produce geography in simple, intelligible, straightforward language they can understand, instead of trying to awe them with incomprehensible jargon and needlessly abstruse mathematical symbolization. Too many geographers seem to labor under the assumption that the work of geographers is somehow inferior to the work of other scholars. In particular, many geographers have been seduced by the aura that is perceived to emanate from Science, and they have assumed that geographers would be held in higher esteem if only they could somehow manage to convince others that geography is a Science. Neither should we be particularly disturbed by the charge that geography is not in the mainstream of the social sciences, because so-called mainstreams in the social sciences have a nasty habit of shifting quite abruptly, and today's mainstream may be tomorrow's stagnant backwater. It was Oscar Wilde, I think, who opined that nothing is outdated more rapidly than that which is most up-to-date.

Scientism in geography seems to stem from two remarkably dissimilar sources, one extremely practical and one that is largely theoretical. The practical source is distribution requirements in the curricula of particular institutions. For example, a course in physical geography satisfies the science requirement in many schools, and it competes quite successfully, sometimes entirely too successfully, with alternative courses. Physicists and chemists would like to see it removed from the list, because they are losing students to it, and the Geography Department must fight tooth and nail to keep it because it is a bread-and-butter course that generates the enrollment necessary for survival. The Department is regularly forced to justify its status among the sciences, and it may be compelled to resort to some ingenious arguments to do so. I will vigorously support these arguments when I am on that particular campus, because I believe quite firmly that a good course in physical geography can contribute just as much, and probably more, to a good liberal education as most courses in physics and chemistry. These arguments should be kept on the campus where they belong, however, and we should have the decency not to inflict on our scholarly peers the kinds of domestic sophistry we are sometimes compelled to practice on deans, curriculum committees, and others of their ilk.

The other source of scientism in geography has been the upheaval, commonly known as the Quantitative Revolution. Our Quantitative Revolution was inevitable, and for the most part it was beneficial, because it opened up the discipline and enabled us to recruit many bright young scholars of mathematical bent who previously might have found geography uncongenial. Some of these new colleagues, unfortunately, brought with them a doctrinaire addiction to scientism. They believed in one particular rigid framework of research design, they had one narrow set of ideas about what was worthy of research, they had the idea that there was only one right way to do things and only one right thing to do, and they were determined to cleanse the temple. Inflamed with the messianic zeal of new converts, they launched a vigorous campaign to purge geography of all that failed to fit their particular definition of Science.

Some members of the new breed have followed the lead of sociologists and deified the idea of "theory." A good theory is nothing more nor less than an idea about the way things may work. It is merely an attempt to impose order and simplicity on complex patterns and processes, and it has little value unless it advances our understanding of the real world.

It is inevitable that I will be misunderstood, so let me try to emphasize and reemphasize the point I wish to make. I am not attacking science. What I am attacking is scientism—the bland assumption that something is automatically superior because someone claims it is scientific, the belief of those who think they will somehow attain higher status and achieve greater prestige if only they can manage to be identified as scientists.

The admirable and essential qualities of good science are equally essential qualities of good scholarship, which is a far larger universe than mere science. Science is subsumed under scholarship, but not vice versa. Our goal as geographers should be high quality scholarship, and we should not fret about whether our good scholarly work happens to fit some particular definition of Science.

Blind Alleys and Dead Ends

The history of geography is littered with the wreckage of attempts to "make it into a Science." These attempts have come a cropper largely because they have arbitrarily tried to amputate some integral part of the discipline, and to exclude some of those who wanted to be known as geographers.

Environmental Determinism

Environmental determinism aroused fierce emotions among geographers a generation ago, but it has been almost completely forgotten today. It began as an attempt to make geography more scientific by replacing teleological explanations with the explanations of natural science. It stemmed from Darwinian notions of natural selection and the survival of the fittest in a given natural environment. The determinists saw the natural environment as cause and human activities as effect. Nature was the physical mold in which human activities were formed and shaped. William Morris Davis wrote: "Any statement is of geographical quality if it contains a reasonable relation of cause and effect between some element of inorganic control and some element of organic response" (Davis, 1906, 71).

The determinists developed a useful standard checklist that included geology, landforms, climate, soils, and vegetation as major elements of the natural environment that were assumed to determine human behavior. They considered natural features so important that they were worthy of study in their own right, but they treated human activities as an afterthought, a mere nuisance, and some geographers even went so far as to propose that human activities should be completely eliminated from geography "on scientific grounds."

The reaction against determinism was led by geographers who realized that people are not merely passive actors on the natural stage, but, as so often happens, they went to extremes, and anyone who even mentioned the environment was immediately accused of determinism. Geographers were intimidated into virtually abandoning the environment well before it had become fashionable once again.

Landscape

Many of those who have wanted to make geography into a Science have labored under the assumption that every self-respecting science must have its own unique objects of study, and they have searched avidly for objects that geography could claim as peculiarly its own. One candidate was the idea of "landscape," a term used by Carl O. Sauer in his inaugural address at the University of California at Berkeley (Sauer, 1925). Sauer was advocating a phenomenological approach in geography, and he was far more interested in morphology than in landscape, which he used as a synonym for area or region, but the would-be scientists eagerly latched on to "landscape" as a basic technical term. They argued that landscapes would provide a unique class of objects for geographers to study, and they expected to be able to make geography into a science by the arbitrary exclusion of all else.

The usefulness of "landscape" as a technical term was diminished somewhat by the fact that no one ever really knows for sure what a landscape is, or is supposed to be, and no geographer has ever been able to define the term to the satisfaction of anyone other than himself. Sauer used "landscape" as the English translation of *Landschaft*, a term that had been widely used by German geographers, but with two quite different meanings. Some of them used it with the common connotations of the things one sees from a given viewpoint, but others used it as a synonym for area or region, and many slid back and forth. They had "woven such an impenetrable web of mysticism" about the term that it could be used to prove almost anything (Hartshorne, 1939, 212).

Some American enthusiasts accepted the colloquial connotation of "landscape," and they insisted that geography could be made into a science by restricting itself to the visible material features of the earth's surface, although this posture forced them to engage in some rather remarkable intellectual gymnastics. At least one of them argued, for example, that there was no geographical justification for distinguishing francophone Canadians from anglophone Canadians because they were not visibly different, and that a geographer was justified in studying the black population of the South only because of the visibility of skin color.

The Science of Regions

The attempt to transform geography into a "science of regions" was based on the notion that regions, rather than the visible landscape, would be ideal objects of study for geographers to claim as peculiarly their own. This brand of "scientific geography" was predicated on two assumptions: first, that regions are real things, "objective realities" that are simply waiting to be discovered if only we search for them with enough diligence and intelligence; and second, that the world is covered with a network of identifiable areas, or "true geographic regions," that are homogeneous with respect to all significant geographic variables. These regions, when identified, would be valid for all purposes and accepted by all geographers. The failure of geographers to identify them was taken merely as a sign of the inadequacies of geographers, and not as a hint that such regions might not exist at all.

Although the idea may seem preposterous to us today, an enormous amount of geographic time and energy was invested in groping for such "true geographic regions." The existence of such regions remained an article of faith for the True Believers, who pursued the quest for them with all the mystical zeal of contemporary model-builders. The long and futile search for "true regions" eventually degenerated into a mere exercise in taxonomy, an attempt to divide up the world or its parts by identifying the areas where the boundaries of significant geographical variables more or less coincided.

Excessive emphasis on the attempt to delimit regions, the sterile "regionalizing ritual," made some geographers forget that the purpose of regional geography is to understand areas, not merely to draw lines around them. Subsequently, a few quantitatively oriented geographers erroneously—but not all surprisingly—assumed that regional geography consisted of little more than the delimitation of regions. They demonstrated that such a trivial exercise in classification was merely a matter of clever computer data manipulation, which missed the whole point: understanding is more important than classification.

Regionalism

The concern of geographers with defining the concept of the region may have been at least partly an attempt to protect their turf against contamination from regionalism, an amorphous and slightly mystical movement that was riding high, wide, and handsome in the 1920s and 30s. Regionalism had little more to do with regions, as they are understood by geographers, than the currently fashionable environmentalism has to do with the natural environment. Regionalism was a shibboleth, a slogan, a code word to identify or to stigmatize those who were opposed to the contemporary world; although there were some notable exceptions, many regionalists seemed to be anti-metropolis, anti-technology, anti-centralization, anti-twentieth century, or, in short, anti-quarian.

The regionalists had turned to the past in their search for order and stability during a period of rapid social change and stress. Regionalism was never a coherent movement, but a hodge-podge of people and ideas. It included nostalgic romantics who felt the urge to live off the land and among the folk in order to regain a sense of identity, and it also included utopian visionaries who hoped to restructure society by redirecting population growth from endless metropolitan sprawl to scattered garden cities protected by green belts.

Regionalism gradually evolved into a political mechanism for extracting a disproportionate share of funds and power from the central government, or for targeting programs designed to redress economic, social, and political inequalities within a nation. The Tennessee Valley Authority (TVA), probably the first and best example in this country, has been followed by the Appalachian and eight other regional commissions and by six river basin commissions. The ad hoc political region provides an attractive basis for economic development because it bypasses the traditional political apparatus, but the "regionalism" that debates alternative strategies of economic development and foreign policy is a far cry from the traditional regional theory of geographers.

No Place for Cities

Regionalism might have been guilty of ant-metropolitan bias, as charged, but the same charge could not have been leveled against traditional regional geography, because it had no stance whatsoever as far as cities were concerned. Regional geographers had focused on the physical and rural aspects of areas, and they were decidedly uncomfortable with cities, which simply did not fit into their scheme of things. Cities are conceptually awkward; they are points, however misshapen, at the continental scale, but they become areas as scale grows larger.

Some geographers realize that each city has its own internal regional geography, but an attempt to translate their ideas about regions from rural areas to the internal structure of cities would have forced them to cope with a new and unfamiliar set of variables. In cities one must deal with environments that are predominantly manmade rather than natural, and the city apparently reduced the value of hard won traditional expertise in geology, landforms, climate, soils, vegetation, and agricultural land use. Careful study of the geography of cities would have exposed the inadequacies of the traditional checklist of regional geography, and many geographers apparently found it easier to ignore cities than to abandon their security blankets.

Furthermore, cities and rural areas require different kinds of mental maps, because rural areas consist of open spaces that are perceived as areas, but spaces in the city are enclosed by buildings, and they are perceived as lines. City travelers see only the facades of buildings, and they may have little sense of what lies inside or behind them. They may have no knowledge of the contents of the areas between their lines of travel. Travelers in rural areas normally can look out across the countryside, and they can get a far better sense of the areas that lie between their routes. The standard maps of cartographers seem to work better in rural areas than in urban areas.

Our Division of Labor

The Quantitative Revolution moved the major emphasis of geography away from regional geography and toward the systematic subfields of the discipline, which are closely related to cognate fields of science. The physical, biotic, and cultural features of the earth have been divided into groups of things that are similar and closely related. The limited set of things in each group is studied by the practitioners of a particular scientific discipline. These things are isolated to permit close examination of the operation of a few critical variables by holding all other variables as constant as possible. The laboratory of the physicist or chemist has served as the model to which other sciences aspire. The practitioners of individual disciplines must develop a form of tunnel vision, and they must restrict their attention to a small number of key variables.

Such narrow concentration is mandatory if one wishes to comprehend the operation of individual variables, but it can have an expensive trade-off if the scholar ignores or only partially understands critical variables that lie beyond the purview of his particular discipline. Any attempt to understand the complex interplay of the myriad variables in the real world requires another kind of approach—the integrative approach, which puts back together, or synthesizes, all of the variables that the individual disciplines have tried to isolate for close analysis. The two disciplines that use the integrative approach are geography and history.

Obviously, no one could hope to be an expert in the enormous range of subject matter that is necessary in an integrative discipline, so geographers and historians must rely heavily on other disciplines for detailed information about the operation of individual variables. There might be no need for systematic geography if all of these cognate disciplines would produce material in the form that is needed by geographers, but in fact they do not, and geography has been compelled to develop systematic subfields to facilitate intellectual intercourse with cognate disciplines. The regional side of geography is concerned with patterns, associations, and synthesis; the systematic side is concerned with analysis of the processes that help us to understand and explain these patterns and associations; and they are two sides of the same coin, geography.

The integrative nature of geography requires its practitioners to be able to wander blithely back and forth across the boundaries between disciplines. One of the great strengths of geographers, in fact, is their ability to cope with a wide range of variables. They cannot pretend to the status of specialist in more than one or two cognate disciplines in the physical sciences, the biological sciences, the social sciences, and the humanities, but any geographer worth his salt has long since lost his amateur standing in all of them.

Why Do We Need Regional Geography?

One of the major consequences of the Quantitative Revolution has been "a questioning of the essential unity and coherence of our subject" (Claval, 1980, 246). The new geography turned its back on the traditional approach of regional synthesis, embraced "the scientific method" and quantitative techniques, and tried to transform geography into a social science in the image of economics. The new approach worked best in the systematic subfields of the discipline. The fledgling geographer was advised to specialize in a systematic subfield, usually the narrower the better, and the generalist and the regional geographer were derided and disparaged as old-fashioned relics of an unworthy past.

We need a unifying theme that ties together all of the wildly disparate phenomena with which we deal, and I know of no other theme that is even remotely so satisfying as the idea of the region. All of the different strands of geography converge when we try to understand a place, an area, or a region, and not one of them can be ignored, which is further evidence of their relevance. We need them all in order to understand regions, and we need the concept of the region in order to understand why we need the diverse and variegated systematic subfields of geography.

The idea of the region provides an integrating philosophical theme for the subject matter of geography, and regions can be useful areas for testing the theories generated by systematic studies, but by far the most compelling argument on behalf of regional geography is that our society expects it of us. Too many geographers have committed the classic blunder of talking when we should have been listening, of telling instead of asking. We make a mistake when we try to tell society what it ought to want from geography, because society already knows what it wants.

Ask whomever you will—the man in the street, the taxpayer, the educated layman, the enlightened citizen, your students, even your colleagues in other departments—what they expect of geography, and to the extent they have any idea at all they will tell you that one of the principal goals of geography has been to satisfy human curiosity about how much of what is where and why it's there, about the where and why of places and people, about the land and how people have used and abused it.

Most people are curious about the world they live in, and they are intrigued by the unique character of particular places. They want to know more than they can see on television or read in newspapers about where and why things are happening, and they realize that they need a framework of systematically organized information about the world in order to understand it. It is inconceivable that anyone could consider himself educated who does not have a basic knowledge of world patterns—the distribution and interrelationships of the major physical, biotic, political, social, and economic lineaments of the contemporary world. Society expects geographers to provide this information in a framework that will facilitate an understanding and appreciation of our large and complex world.

Regional geography provides such a framework. The region is basically a pedagogic device for organizing information for the most effective presentation, comprehension, and retention. It is a truly marvelous pedagogic device, not in the narrow and restricted sense of what transpires within the four walls of a classroom, but in the far grander sense of communicating information, exciting enthusiasm, and stimulating curiosity. Pedagogy is an art, and great teaching, like great art, should be uplifting and inspiring as well as enlightening.

When I say that the region is basically a pedagogic device, I do not mean to imply that regional geography can exist without first-rate original research, because we cannot accept the widespread myth that good teaching can be divorced from good research. Research and teaching are inseparable partners in the process of tightly disciplined curiosity and communication that is known as scholarship. A true scholar combines an insatiable curiosity with an overwhelming compulsion to communicate the knowledge and insights he has acquired, whether to a narrow circle of listeners or to the wider audience made available by the printed page. Just as research without publication is sterile, so teaching that is not grounded on research is a lifeless mockery, the mere transmission of second-hand ideas and knowledge that lacks immediacy and excitement.

Good Regional Geography

No one has improved on the principle enunciated by Nevin Fenneman in the classic paper in which he defined the physiographic provinces of the United States: the best region is the one that permits the largest possible number of general statements before details and exceptions become necessary (Fenneman, 1914, 87). Spate said essentially the same thing: the best region is the one that satisfies the greatest number of significant criteria while doing the least violence to any of them (Spate and Learmonth, 1967, 410).

Good regional geography should begin with the dominant theme of each region, which of course will vary from region to region. No standard list of criteria or checklist of features-to-be-observed can be universally applicable to the study of all regions, although three key themes will recur in nearly every region.

First, regional geography must be informed by a sense of time, or becomingness. You cannot understand a region until you understand how it came to be the way it is, and how it is changing. Present patterns are but a momentary reflection of continuing processes of change, and we must be sensitive to the direction and rate of change, both past and future.

The first event, which may well have been random, has far greater influence on subsequent developments than we sometimes realize. The first raindrops that furrow a newly emergent slope will affect the entire subsequent history of landform evolution; the first seedlings that sprout on newly bared ground will create an environment that is hospitable to some plants and hostile to others; the first houses and trails in an area of new settlement will influence all later developments, and it is extraordinarily difficult to make major changes in the basic street pattern of a city.

A second key theme in regional geography is the relationship between the size of regions, scale, and the amount of detail that is appropriate at different scales. There are no preordained rules that stipulate the proper size of regions. The scale of a regional study will determine whether a particular part of the earth's surface is treated as a point or as an area. Even our largest cities, for example, are treated as points on the map of the United States, but cities contain enormous internal diversity, and they are well worthy of regional description in their own right.

Greater New York City certainly is quite as important as most of the other regions that are treated in our standard regional textbooks, but it has been virtually ignored in the literature of geography. Surely Wall Street, the lower East Side, Seventh Avenue, Midtown, Harlem, Westchester County, Long Island, and the like are just as much a part of the geography of the United States as the Bluegrass, the Badlands, and the Blackland Prairie.

A third key theme in regional geography is the importance of the physical environment, which provides the stuff that people have to work with, and sets the stage on which the human drama is played. This stage has been variously modeled and remodeled by different groups of people, and at times they have been tempted to play fast and loose with it, but old Mother Nature is always watching, and people ignore her at their own very great risk. The physical environment is important even in cities, where it has most often been ignored. The traditional heavy orientation of physical geography toward rural areas requires rethinking. The length of the growing season is less important in cities than heating and cooling requirements, solar energy potentials, or hazardous driving conditions. Contemporary Americans are city people, and they need to know more about the use of rivers and the atmosphere as sewers, about municipal water supply, about soil stability and permeability, about the relationship between glacial topography and basement flooding.

Any geographer who takes leave of his armchair, and gets up and goes out and takes a close look at the world around him, is going to be forced to conclude that Mother Nature still plays a major role in human affairs. Geographers, like the mythical giant Anteus, derive their strength from contact with the earth. Anteus became stronger each time he was hurled to the ground, and Hercules finally was able to vanquish him only by holding him aloft. Perhaps we should designate Anteus as one of the patron saints of geography, because we too grow weaker the farther we depart from Mother Earth.

The Necessity of Fieldwork

Our other patron saint certainly would be Argus, the man of a hundred eyes, because good regional geography, and I suspect most good geography of any stripe, begins by looking. We start with the things we see, the visible features of the earth's surface—the form of the land, the vegetation that cloaks it, and the structures that people have added. Our search for understanding quickly transcends these visible physical features, however, and we attempt to get inside the minds of people, because we cannot be satisfied until we understand the values that motivate their behavior.

Unlike scholars in other disciplines, who can restrict their attention to a limited number of variables, geographers are interested in the totality of places, and we need to see how apparently disparate things fit together in the real world. We cannot understand an area until we have explored it for ourselves. Of course, we read what others have written about it, but we remain uneasy with our knowledge until we have had a chance to test it in the field.

We visit new areas with keen anticipation of what we will find, and we are excited when we discover the unexpected, because it makes us generate new and better explanatory hypotheses. We return to familiar areas alert for changes; our greater familiarity enables us to notice nuances we had not previously observed and to develop even subtler hypotheses. We are lost without maps—and sometimes we are lost even when we have them. They are our basic tools for recording and analyzing data. Field observation and cartographic analysis generate our most geographical hypotheses, which we return to the field to test.

One of the most important field techniques in contemporary geography is talking to people. We need to talk to all kinds of people, on their own turf and at their own level, whatever that level may be, to find out what they think, why and how they do things, and what values they hold dear. We cannot understand a region until we understand the behavior of the people who inhabit it, and we cannot understand their behavior until we understand and appreciate the values that motivate that behavior.

The Importance of Values

Value systems differ from person to person, from group to group, from region to region, and so does rational behavior based on those different value systems.

People in the inner city really are different from those in the suburbs, and people in South Carolina really are different from people in South Dakota. They talk differently, they dress differently, they think differently, and they will behave differently when they are confronted with identical circumstances, because they are motivated by different value systems. Regional geography can make an important contribution to a liberal education by helping students to realize that different groups of people in different places really are different.

Cultural geographers, who should have been able to make a special contribution to our understanding and appreciation of the importance of noneconomic values, have let us down rather badly, at least in the United States, because their approach to values has been so unimaginative. They have focused on those groups that "have worn name tags," at the expense of groups that have not been easily identifiable by such obvious distinctions as country of birth, language, and religious affiliation, and they have largely ignored the importance of region of birth, social class, and other more subtle distinctions.

Furthermore, all too often their approach has been completely unscholarly. "Here is an area that was settled by Germans," they have said. "Let's find out what is distinctively German about it." Such behavior is a bit like shooting fish in a barrel that has no bottom. It would be better to begin by identifying differences—using them as the dependent variable, if I may be permitted a big of jargon—and then seeking the cultural variables that might help to explain these differences.

A *Lifelong Commitment*

There are no instant experts in regional geography. An attempt to understand any region must be based on sensitive interpretation of the values that motivate the behavior of its people. Such an interpretation is a lifetime pursuit that requires prolonged personal experience of the region and those who live in it. The study of a region must be such a lifetime commitment that a geographer should choose his or her region with great care. You should select a region in which you are so interested that you are willing to dedicate a major part of your life to getting to know it intimately. It is a good idea to capitalize on your experience, and to choose a region in which you were born and raised, or one in which you have lived for some time, but we also need geographers who are courageous enough to make commitments to the study of foreign areas.

The regional geographer must get to know his own region well, but he must not pass up opportunities to get to know other areas also, because he needs some basis for comparison. If I may use one of my favorite bits of geographically apposite doggerel,

> He knows naught of England
> Who only England knows.

One of the more serious weaknesses of contemporary American geographers may be their lack of field experience in foreign areas, because they cannot understand the distinctiveness of their own part of the world until they know how it differs from other parts.

Effective Communication

The understanding, appreciation, and interpretation of any part of the world is completely worthless until the insights one has gained have been effectively communicated to others. Communication is the sole purpose of scholarship. The uncorrupted epigram is, "Publish your results, or they will perish with you." A scholar may know all there is to know about some particular subject, or about some particular part of the world, but his knowledge does not exist until he has communicated it to others. Far and away the most effective and appropriate medium of communication is the printed page, with lectures a distant second. A scholar must learn how to write clearly and effectively. The goal of good writing is to express profound ideas so clearly that they seem simple; poor writers say simple things in a way that makes them sound profound. In writing, as in all else, the hard part is making it look easy.

Good geographical writing quite definitely is an art, and the highest form of the geographer's art is writing evocative descriptions that facilitate an understanding and an appreciation of places, areas, and regions. Good regional geography should be written with imagination, flair, style, verve, dash, panache, enthusiasm, vivacity, animation, and perhaps even a bit of flamboyance. Above all, it should be fun to read. Writing imaginative, informative, and readable descriptions of places is an extremely difficult of accomplishment, and it demands a creative flair that is as much intuitive and emotional as intellectual.

Writing good regional geography is a personal artistic effort, and a touch of genius obviously is helpful, as it is in any endeavor, but I would remind you of the race between the tortoise and the hare. Those of us who are willing to work hard may be able to make contributions nearly as great as those that are made by our colleagues who are far more richly endowed. Few of us can aspire to greatness as painters or musicians or poets, but our lack of talent need not deter us from enjoying music and poetry, or even from trying to produce a bit of it after our own modest fashion. Sensitivity may be almost as important as creativity in the production of good regional geography.

Regional geographers must learn to believe what their data tell them, instead of trying to force the facts into a mold of preconceived theory. They must realize that conventional wisdom often is out of date, and the only way to understand the world is to get out and explore it with sensitivity and with sympathy. The good scholar must learn to look at the real world, in addition to documentary sources. He must learn to talk to ordinary people, not just to other scholars, and above all he must learn to listen, and to listen carefully, if he wishes to understand, and to help others understand, this fascinating world.

I am not about to break my own rule by insisting that all geographers drop everything else and rush out to become regional geographers, because geography has entirely too many important jobs to do, and far too few competent and qualified people to do them, but I do hope that all geographers will recognize the crucial importance of regional geography to the discipline, and that at least a few will accept my challenge and my invitation to adopt a region, to immerse themselves in its culture, to acquire a specialist understanding of it, and to contribute to our store of knowledge about some small segment of this complex, surprising, exciting, and utterly magnificent world we live in.

Reading

CHAPTER 2

the Landscape*

Queen Elizabeth I of England once said that she was "mere English." I cheerfully admit that I am content to be a mere geographer, and I wish to claim no more. I am a geographer because I am fascinated by the complex diversity of the earth we inhabit. In simple terms, I might define geography as curiosity about places. Geographers have both the privilege and the duty of trying to understand and explain the distinctive character of places. The inevitable starting point for geography is the visible landscape, because it is the most obvious feature of any place, although of course we transcend it almost immediately in our search for explanations.

I am quite content with the simple vernacular definition of landscape as "the things we see," and I am saddened by the way the meaning of the word has been transmogrified. Some people have endowed the concept of landscape with magical, mystical, or symbolic significance, or have loaded it with metaphysical connotation. In some quarters the word even has been turned into a talisman, as though it were expected to bring good luck to anyone who uttered it.

Many users have merely purloined part of the word. They have assumed, quite incorrectly in my opinion, that the suffix –scape in landscape is somehow related to the suffix –scope (as in microscope and telescope), and they have committed such awkward neologisms as cityscape, farmscape, roadscape, seascape, wildscape, windowscape, and the like(scape).

* "Reading the Landscape," Chapter 3 in George F. Thompson, editor, *Landscape in America* (Austin: The University of Texas Press, 995), 23–42.

Some users have treated a single facet of the landscape as though it were the whole. Geomorphologists, for example, have written about "landscape evolution" when they were concerned only with rocks and the features of the land surface. Some botanists and ecologists have used the word when they were interested primarily in plants and vegetation. The landscape of the architect seems to consist mainly of buildings, and often only of those buildings that are "polite."

Some people have used the word metaphorically, with results that can be confusing. The "economic landscape" apparently is no more than a flat, uniform surface on which only markets matter, such as Merrill Lynch's "landscape of investment," but what is a "political landscape," or a "demographic landscape," or the "landscape of shirt features" advertised in the Lands' End catalog, or the "landscape of TV" the Associated Press natters about? The mind boggles at the very thought of a "soil landscape," which presumably would be of interest only to another earthworm.

I have already said I am content with the simple vernacular definition of landscape as "the things we see." Our perceptions may differ, but I doubt they differ all that much. You and I both see the same thing when we look at a skyscraper, or an elm tree, or the rear end of a bull. Our skills at reading what we see may differ, to be sure, and what we see may well say different things to us. One person, for example, might see picturesque rocky cliffs, whereas another might see the Niagara dolomite outcropping in a cuestaform escarpment, and both would be right. But a few of the Deep Thinkers like to inform us that what we see is not what we see at all, and that reality is not really real. Hogwash.

We need to get out and look and see things for ourselves, because the primary source for studies of the landscape, oddly enough, is the landscape itself. I do not know how one learns to become a careful observer, so I am unable to teach the skill, but I know it is essential.

We also need to learn to think analytically about what we have seen, and to identify the features of the landscape in terms that make sense to, and can be used by, others. Too many studies of landscape have to be taken on faith, because their authors cannot explain the criteria they used to identify the features they have described.

We need to learn to talk to, listen to, and learn from the plain ordinary people who are the creators, the inhabitants, and the custodians of the landscape. We must know enough to ask reasonably intelligent questions, and, of even greater importance, to listen carefully to what they tell us. An old adage says that one is not going to learn very much while one's mouth is open.

Good fieldwork requires an open-ended kind of approach. In order to glean the occasional gem, one has to be willing to take the time to relax, to chew the fat and shoot the bull, to listen patiently to a lot of things in which one is not interested. In my time I have listened to far more homilies on politics and religion than I have really needed, but I have also learned a lot because I have been able to hear them out attentively, and then turn the conversation to subjects that interest me more.

We can discover the aspirations, needs, and values of ordinary people only by listening to them, because they rarely, reluctantly, and with painful effort put pen to paper. One can have little confidence in the work of secondhanders who seldom can be bothered to glance out of their windows, whose experience has been limited to the dusty stacks in the library, and who are hostage to the observations and ideas of others because they rely entirely on the printed page.

The printed page undeniably is important. It enables us to visit unfamiliar landscapes vicariously, and to learn how others have seen and interpreted landscapes with which we are unfamiliar, but far too many of us are far too willing to let others think for us. We must learn to believe what we see and to have confidence in our own observations, especially when they fly in the face of conventional wisdom and generally accepted theory. Most of us are intellectual cowards, and it is far easier to defer to authority than to risk sticking our necks out. We must develop the courage to believe the evidence of our senses, and to let the facts speak for themselves, instead of trying to force our observations into some procrustean model of theory or conventional wisdom.

The observer of landscape should try to see everything, but most landscapes and their histories are so complex and so variegated that they are virtually incomprehensible without careful analysis, and the person who tries to look at everything may wind up seeing nothing at all. We must be selective. At any given time we must concentrate on a few carefully chosen features, or types of features, and we must never allow ourselves to forget that the features on which we concentrate are related in various ways, some close and some not so close, to all the other features of the landscape.

It is logical, convenient, and defensible to group the features of the landscape into three broad categories—animal, vegetable, and mineral—to facilitate their analysis. The three principal components of any landscape are:

(1) the landforms, or the features of the land surface;

(2) the vegetation, or the plants that cover the surface; and
(3) the structures that have been added by people.

The human structures may be further subdivided into four categories:

(3a) systems of land division;
(3b) structures associated with the economy;
(3c) house types; and
(3d) agglomerations of houses into villages, towns, and cities, with all of their associated features.

In most parts of the world, the aspect of the landscape that makes the most compelling first impression is the form of the land surface, whether mountains, hills, plains, or plateaus. The observer of landscape must have a good command of the basic principles of geomorphology in order to comprehend the character and origin of the major and minor forms of the earth's surface. In addition to their role as features of the landscape, the landforms of an area usually are closely related to the kinds of plants that grow in the area and to the ways in which people can use the land.

The second principal component of the landscape is the plants that cloak the surface of the earth. The observer of landscape should be familiar with such concepts as natural vegetation, contact vegetation, climax vegetation, potential vegetation, and the problems inherent in each concept, but he or she is primarily interested in the real vegetation, which can be defined as the plants that are there now.

The cover of trees, shrubs, grasses, and other plants is also an essential intermediary between the mineral and animal worlds. It is trite, but true, to say that animals (including the human variety) cannot eat rocks, and they would perish if they did not have plants to convert the minerals of the earth's crust into food they can use. Animals are mobile, and they are free to wander, but plants are sedentary, and they are firmly rooted in the soil. Animals cannot live or roam too far from the places where the plants they need are available, or where they can be grown.

All plants need heat, light, and water. The proper amounts vary from species to species. The requirements of the more important plant species provide the basis for most classifications of climate, and the map of climatic regions looks very much like the map of vegetation zones. Within each climatic region, however, the vegetation varies in response to differences in elevation, slope, exposure, drainage, soil, and other environmental factors.

The plants that grow naturally in an area probably are the best single indicator of its environmental potential, and pioneers in the United States used vegetation as a guide to tell them which land to settle and which land to avoid. Their knowledge is embedded in folk sayings, such as, "Maple land is good land, but pine land is poor land." Even today, the observer of landscape should know how to read the vegetation, because it carries a clear message about the nature of the environment in which it grows.

The vegetation also tells the tale of how people have used and abused the land. For example, most of the eastern United States was wooded when white settlers first saw it, but they cut down the trees and cleared the land so they could grow crops on it. Much of the land still remains in cultivation, but some was cleared unwisely, and those who once farmed it have allowed Old Mother Nature to foreclose her mortgage on it. They may be reluctant to admit they are no longer using the land effectively, but broomsedge, blackberry bushes, persimmon sprouts, cedar saplings, old field pines, and other plants send a clear signal that the land is no longer used for agriculture. The observer of landscape should be able to recognize the plants that invade and colonize unused agricultural land.

We should also be able to identify the principal cultivated crops, and should understand why they are grown and how they are used. A generation ago, for instance, a traveler in the Corn Belt of the American Middle West might have expected to see small, well-fenced fields of corn, oats, and alfalfa, which were grown in a regular rotation to provide feed for cattle and hogs. Today, however, the Middle West seems to have been turned into one vast field of corn and soybeans, both of which are sold for cash instead of being fed to livestock. Fences have been removed, fields have been enlarged, farmsteads have been transformed, and the entire rural landscape has been changed in response to changes in agricultural technology and economy.

The fences, fields, farmsteads, and other structures added by people are the third principal component of the landscape. Every human structure serves some human need, and the form of the structure fairly faithfully reflects its function, or the need it was originally designed to serve. Form and function are more intimately related in some types of structures than in others. They are most closely related in simple, ordinary, workaday structures, such as barns, sheds, garages, and privies, that are necessary for daily living; they are most widely divorced in expensive, pretentious, ornamental, showplace structures, such as houses, and that is why the study of house types is so extraordinarily complex, difficult, and usually unsatisfactory.

I prefer to invoke function to explain as much of the built landscape as possible, but, of course, the form and appearance of the human structures in any area may also be influenced by many other considerations, including the nature of the physical environment, the level of economic prosperity, the technical competence of the builders, the date of construction, the cultural baggage of the people, aesthetic and symbolic motives, and the personal whims and idiosyncrasies of the owner.

Ordinary vernacular structures normally are built of whatever material comes most readily to hand, and, as a consequence, they often reflect the natural physical environment, which provides the stuff that people have to work with. Take fences, for example. In the eastern United States, which was a wooded area, most of the early settlers put up zigzag fences of rails split from trees they had cleared from the land, but stone walls are more common in the stony, glaciated areas of New England and in areas where the limestone bedrock formed natural flagstones. On the sandy soils of the cutover lands of northern Michigan, Minnesota, and Wisconsin, farmers uprooted the tree stumps left by lumbermen and tipped them on their sides to make picturesque stump fences.

The treeless prairies of the continental interior were not as much of a deterrent to settlement as some people have supposed, but settlers who had come from wooded areas did have to scratch around a bit to find substitutes for wood for fuel, for building materials, and for fencing. They experimented with live hedgerows before the invention of barbed wire facilitated the expansion of settlement westward across the grasslands.

The availability of barbed wire, which was made in distant factories and had to be shipped to the grasslands at no small expense, suggests how the level of economic prosperity in an area can influence the form and appearance of its human structures. Building materials for ordinary structures are not normally transported any great distance, but a prosperous area can afford to ship in materials that a less prosperous area could not afford. Steady improvements in the technology of transport also have enabled many areas to import building materials from ever greater distances, and in the United States today few distinctive building materials are derived directly from the resources of the local physical environment.

The technical competence of the people who build the structures can also influence their form and appearance. Barn roofs illustrate the importance of technical skill. Almost any good carpenter can put up a straight gable roof, but straight roofs may be inefficient, because loft space under the eaves is cramped, and it requires much bending, stooping, and head-bumping. Gambrel roofs, with gentle upper sections and steep lower sections, permit easier access to the space under the eaves, but building them demands greater skill. Arched (gothic) roofs, which are even more efficient, are also the most difficult to build. It appears that most barns in the Middle West in the nineteenth century had straight roofs, but gambrel roofs gradually became more common as carpenters learned how to build them, and many barns built since the Depression have had arched roofs, whose principal structural members are bought prefabricated at the local lumber yard.

The date of construction, therefore, may affect the form and appearance of a building, both because improvements in technology over the years have given builders greater leeway and because the style of a building may reflect the changing fads and fancies of current fashion. House types are a good illustration. In the American colonies before the Revolutionary War, buildings were patterned after models in the home country, and the austere elegance of the formal Georgian style was high fashion. After the Revolution, the young republic turned for its model to ancient Greece, the cradle of democracy, and the white pillars and porticoes of Greek temples were copied in the Greek Revival style in the latter part of the nineteenth century, only to be replaced in turn by the California bungalow of the interwar years and the no-nonsense functionalism of the ranch style and split-level ramblers since World War II.

The cultural baggage of a group may affect the form and appearance of its structures, because different groups of people have their own distinctive ideas about how particular types of buildings should look. Some cultural geographers, in fact, have argued that house types are among the best indicators of cultural diffusion. A good example of a type of building that is associated with a particular cultural group is the Pennsylvania German barn, a massive two-level structure with a barn bank, or ramp, on one side and a forebay sticking out over the stockyard on the other. German farmers from southeastern Pennsylvania carried the idea of this barn with them when they migrated westward, and they built such barns wherever they settled. The Pennsylvania German barn was unnecessarily elaborate for the farming system that developed in the Corn Belt, however, and most non-German farmers in the Middle West preferred to build simpler feeder barns of the type that had originated in the Great Valley of northeastern Tennessee, but German farmers persisted in building barns of the familiar type.

Aesthetic and symbolic considerations undoubtedly affect the form and appearance of many human structures, but it is easy to assign too much importance to them. Most people are motivated by functional, not aesthetic, considerations when they erect a structure, and most ordinary human structures must be understood in terms of their functions. They are not intended as works of art, and any artistic quality they may happen to possess is unconscious, accidental, and incidental. People do not intentionally erect structures that are ugly, of course, but neither do they erect structures because they are beautiful; they erect them because they are needed.

Most of us would like to live in a fine house, drive a fancy car, and wear the latest fashions, but few of us can afford the luxury of style, and the way we actually live is a compromise between our desires and our income. We like beauty—even though our perceptions of what is beautiful may not agree with the perceptions of others—but we are compelled to balance our desire for beauty against the dictates of our pocketbooks. Each of us harbors notions of rightness and propriety about the way things ought to be, and we try to implement our ideals, but we cannot always afford to do so. We do the best we can with what we have, and the results sometimes are quite handsome, but sometimes less so, and not everyone agrees about what is handsome. Some people, for example, like statues of pink flamingos, deer, and jockey boys in their front yards, whereas others prefer boxwood hedges, carpets of juniper, and brick walkways, and each is aghast at the awful taste of the other group.

We must try to understand the human structures on the landscape in terms of their creators, inhabitants, and custodians, and in terms of the things with which these people are comfortable. For example, take Clarence Magnus's farmstead southeast of Slayton, Minnesota. The two-story white farmhouse sits at the back of a neatly manicured lawn, with enough trees to provide welcome relief from the scorching summer sun. Gleaming metal sheds and grain bins encircle the clean gravel farmyard back of the house. At the far corner, next to three cylindrical concrete silos, is a handsome old barn from which the white paint has started to fleck.

Southeast of the farmhouse, and hidden from it by a high hedge, but clearly visible from the highway, is a tatterdemalion collection of disc plows, chisel plows, field cultivators, stalk choppers, grain augers, wagons, and other equipment, all overgrown with weeds and apparently rusting away. At first glance, it looks like a scrap-metal dealer's paradise, but when you take a closer look it is not nearly as unkempt and disorganized as it appears to the motorist speeding past on the highway.

Clarence is a splendid manager, and he carefully oiled each piece of equipment and gave it a protective coat of grease before he put it in the lot. His existing buildings simply do not have enough room to house all of the new machinery he needs to conduct a modern farm operation, and he has to leave some of his equipment outside, mostly the cheaper and hardier pieces. Does he need a new machine shed? Of course he does, and he will tell you so himself. He will also tell you, to the penny, what it will cost, including interest and the increase in his taxes because of what the new machine shed will do to the assessed valuation of his farm.

Clarence needs a new machine shed, but right now he can't afford it. One of these years he is going to get a good price for his corn or his beans or his beef cattle, and that's when he is going to build himself one. It will be a hulking metal affair, probably gleaming white with green trim. Aesthetes will consider it an eyesore and Clarence quite frankly doesn't give a hoot whether some college professor or historic preservation officer happens to consider it an eyesore, because it is going to make his farm more efficient and his life easier.

Fred Kniffen (1900–1993), the cultural geographer, once remarked that no one is ever going to wax sentimental over those ugly metal structures that are popping up all over the place. I disagreed with him then, and I still do. It probably will not happen in our lifetimes, and it is unlikely to happen until they have become old and rare, but eventually they will become old and rare, and then they will become precious. Everyone laughed at me back in the 1940s when I predicted we would live to see the formation of a Society for the Preservation of Victorian Monstrosities. Today everyone is far too busy snapping them up and stripping them and refinishing them and gentrifying them to have time to laugh at me any more. I am prepared to make the same prediction for Clarence Magnus's new metal machine shed, the one that hasn't even been built. One of these days it will be added to the National Register of Historic Places, but, of course, neither you nor I will be around long enough to find out whether my prediction will come true.

Different people perceive and assess the same landscape quite differently, and praiseworthy attempts have been made to measure differences in landscape assessment as objectively as possible, but the principal conclusion that emerges from such attempts is the complexity of the problem. We human beings are a cantankerous lot, and we seem to become increasingly cantankerous as we get farther from the constraints of having to make a living, and as we have greater freedom to indulge our personal whims and idiosyncrasies.

The form and appearance of human structures on the landscape are influenced by so many different considerations, from personal eccentricities to the necessities of making a living, that they have attracted the interest of scholars from many different disciplines. These scholars take different approaches to the study of landscape. They begin with different basic assumptions, they emphasize different aspects of structures (or whatever else they are investigating), they collect different kinds of evidence, and they lean toward different explanations. Although our approaches are different, each of us has an important job to do and a useful contribution to make, and we need to work together, in a spirit of mutual appreciation and respect, toward the common goal of a richer and fuller understanding of the landscapes that intrigue us all.

The

CHAPTER 3

British Moorlands*

The unimproved pastures of the British Moorlands traditionally have been used for breeding hardy cattle and semi-wild hill sheep, which are sold to lowland farmers for fattening. To a lesser degree, the moorlands have also been the haunts of sportsmen and vacationers from the lowlands. In recent years, schemes have been put forward for improving the vegetation of the moorlands, and thus increasing their grazing value. Large moorland acreages also have been afforested with coniferous plantations. It is apparent that many, and sometimes conflicting, interests are focused on making more effective use of the British moorlands. An intelligent evaluation of their claims, and thus of the optimum utilization of the moorlands, must be based upon an understanding of the moorland environment and of its adaptability to various uses.

Area, Location, and Use of the British Moorlands

Variations within and between moorland areas make the term *moorland* difficult to define precisely. As a general rule, a moorland is at considerable elevation—commonly more than eight hundred feet above sea level—and is characterized by excessive moisture, peat soils, and unimproved, semi-natural vegetation dominated by heather and coarse grasses. Moorlands are frequently referred to as *mountain and hill lands*, *hill lands*, or *hill pastures*, because of their elevation, and equally often they are called *rough grazings* because of their dominant use. Although *moorland* and *rough grazing land* are generally

---

* From *The British Moorlands: A Problem in Land Utilization* (Athens: University of Georgia Press, 1955), 1–46.

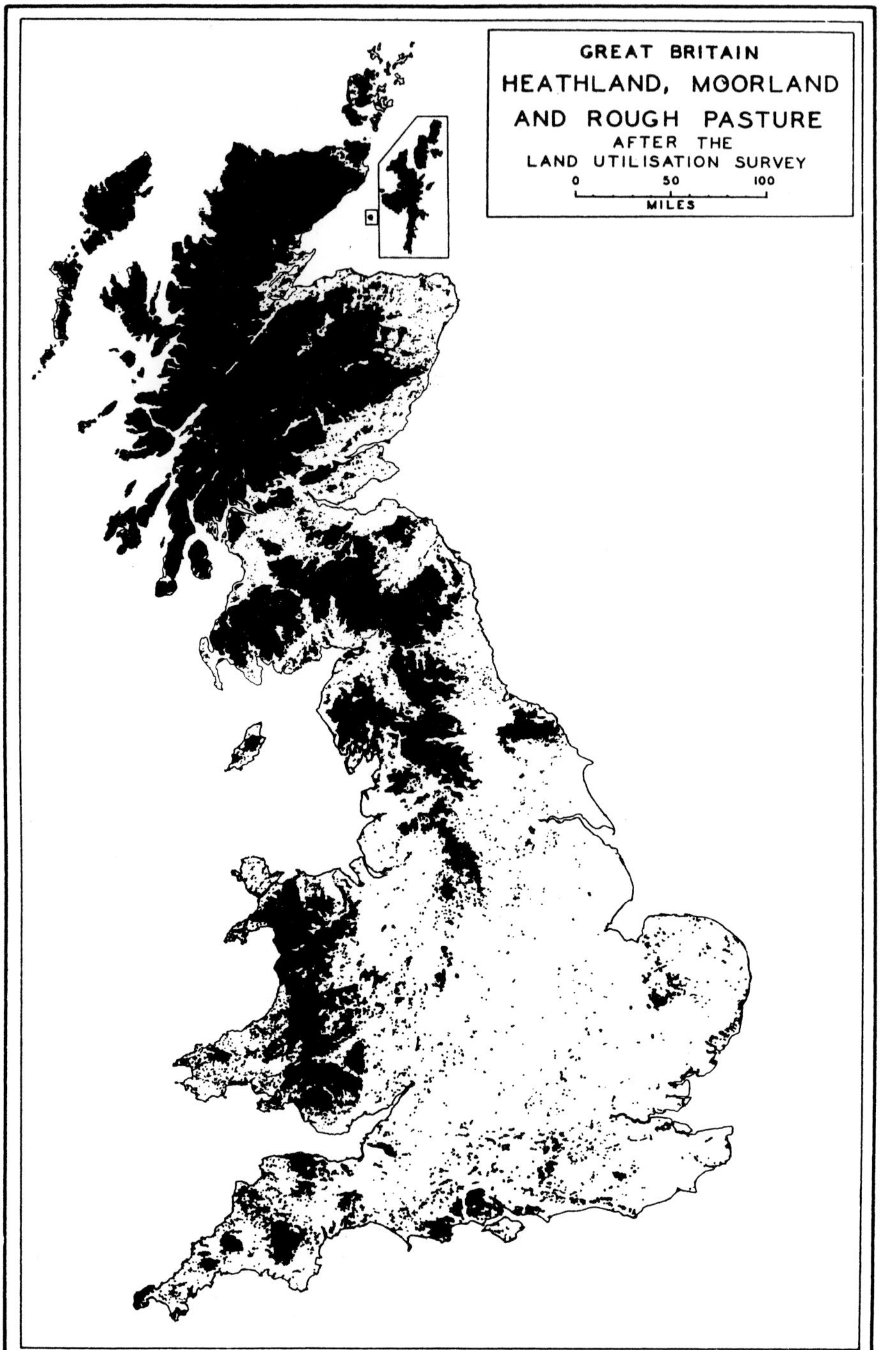

FIG. 3.1. Heathland, moorland, and rough pasture in Great Britain.

synonymous, some moorlands are insufficiently pastured to be considered rough grazings, whereas others reserved for sporting purposes are called *deer forests* or *grouse moors*. In similar fashion, some rough grazing areas, such as the dry, sandy *heaths* of lowland areas in southeastern England, cannot properly be considered moorlands. It is evident, on the other hand, that at least nine-tenths of the rough grazings of Britain are moorlands and, conversely, that at least nine-tenths of the moorlands are used for rough grazing.

The Land Utilization Survey of Britain, which was carried out between 1931 and 1939, provides the best available information as to the actual location and extent of moorlands in Britain, although moorlands were unfortunately grouped with heathlands for the purposes of the Survey. The Land Utilization Survey revealed that 29,266 square miles of Britain were used as "heathland, moorland, commons, and rough hill pasture," and indicated that almost ninety percent (25,778 square miles) of such land was moorland. For a comparable period, the Agricultural Census, taking cognizance only of land used for grazing, reported an average of 24,600 square miles of rough grazing land, but failed to account for 16,800 square miles of land which were not in agricultural use. There seems little reason to doubt that at least 25,000 square miles, or almost a third of the island, is moorland.

The area of moorland in Britain has been slowly but steadily increasing, both absolutely and relatively, during the last sixty years. Between 1892 and 1951 the total area of land reported as agriculturally productive has averaged slightly more than 70,000 square miles. Rough grazings totaled little more than a quarter of this area in 1892, but now account for almost two-fifths of it. This striking increase in moorlands is due primarily to widespread abandonment of marginal crop and pasture land. The major moorland cores have remained virtually untouched by crop cultivation while their fringe areas have expanded into abandoned fields and pastures.

The four major moorland areas in Britain are: (a) the Highlands of Scotland, (b) the uplands of southern Scotland, the Cheviot Hills, the Lake Districts, the Pennines, and the Peak District, with an outlier in the Cleveland Hills, (c) the uplands of Wales, and (d) the uplands of the Southwest (Fig. 3.1). These large blocks of moorland are associated with the hill masses of Highland Britain, which are underlain by resistant Paleozoic and crystalline rocks.

The farmers of Lowland Britain, like the city dwellers, have close ties with the moorlands. British agriculture is dominated by livestock, and moorland

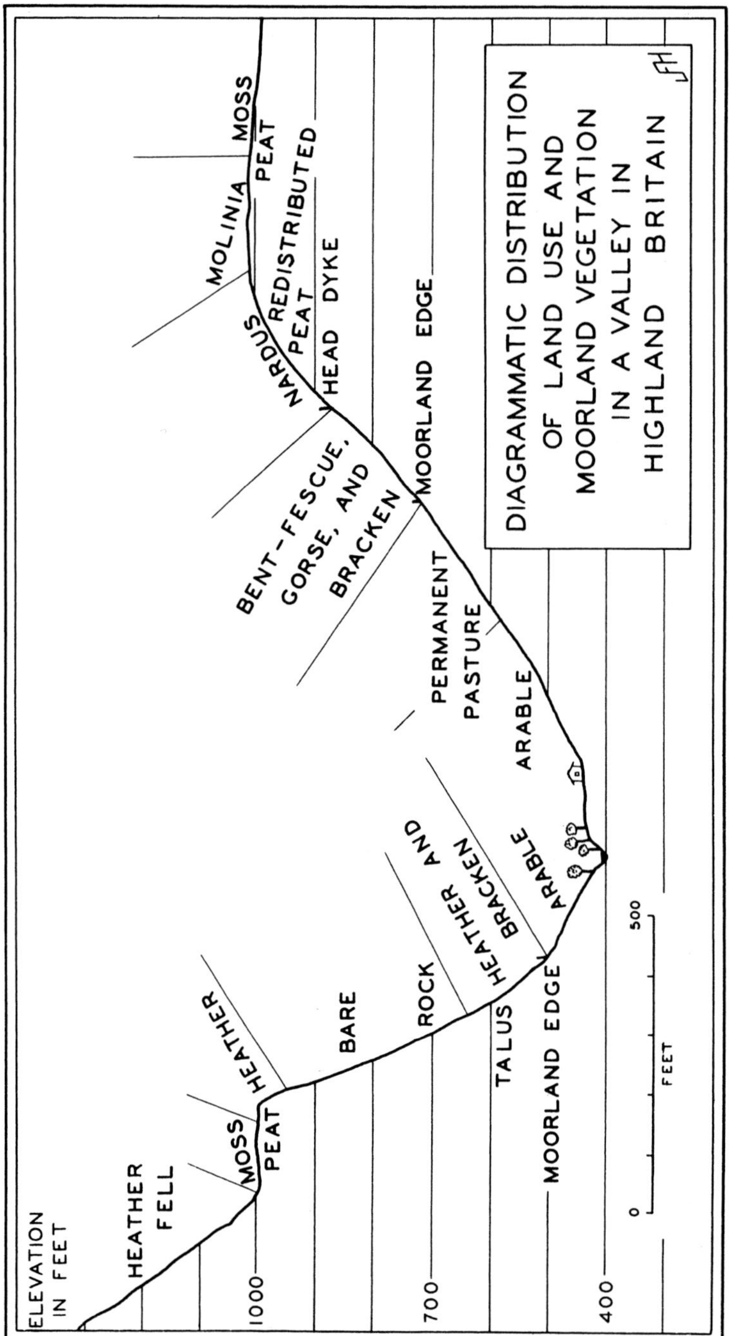

FIG. 3.2. Land use and moorland vegetation in a valley in highland Britain.

farms have a great reservoir of hardy breeding stock from which lowland herds are replenished. The movement of animals from moorland to lowland fattening farms takes place in several stages and over several generations, but many herds of fattening livestock in Lowland Britain may be traced back to breeding herds in the moorlands.

The Moorland Environment

A moorland remains in a natural state because it cannot be improved economically either for grazing or for cultivation. The upper limit of improved land—or the *moorland edge*, to use the common name for the boundary between improved land and moorland—is closely related to the natural environment throughout Britain. Relief, elevation, precipitation, soil development, and other factors of the natural environment coincide with the moorland edge. Even though no specific causal relationships are postulated, there is ample reason for believing that these physical factors play an important, if not a dominant, role in condemning much of Highland Britain to moorland.

Except for the southwestern peninsula, the relief of Highland Britain is primarily the result of the severe glacial erosion to which its crystalline rocks were subjected during the Pleistocene Epoch. Moving ice gashed deep, steep-walled valleys into a relatively level upland surface, and scraped away most of the soil of the uplands. Ice also gouged many higher summits into jagged peaks. Material scoured from the uplands was plastered over adjacent lowland areas or deposited in the lower and wider valleys. Highland Britain, in consequence, has four characteristic surface forms: (a) rugged ice-scoured *peaks*, rising above (b) level to rolling *uplands*, which in turn are gashed by (c) deep, steep-sided *valleys* that widen into (d) drift-covered *lowlands*. Arable land is largely restricted to the drier lowlands and more protected valleys. Gentle protected slopes below about eight hundred feet are used for permanent pastures of improved grassland, but steep, ice-scoured slopes are left as bare rock or unimproved moorland (Fig. 3.2).

The cumulative effects of lower temperatures, heavier precipitation, and stronger winds place stringent limitations on the use of land more than eight hundred feet above sea level. Throughout most of Britain the upper economic limit of cultivation commonly lies between the seven and eight hundred foot contours, although improved pastures and even cultivated fields are found at altitudes greater than a thousand feet in protected areas. There is an obvious correlation between moorland and land more than eight hundred feet above sea level, and exceptions are primarily the result of differences in latitude and exposure. Warmer temperatures in the south permit agricultural use of areas up to a thousand feet above sea level, whereas large moorland areas in northern Scotland are at elevations of four hundred feet or lower. Great exposure to wind and rain likewise restrict agricultural use of the western parts of upland areas. Moorland is actually found at sea level along the west coast of Scotland.

In addition to the handicaps imposed by relief, altitude, and temperature, the British moorlands are sodden. Heavy precipitation and low evaporation rates produce an excess of moisture; thus, the majority of moorland areas are inadequately drained. Stamp (1948) estimates that grain production in Britain is uncertain where precipitation exceeds thirty inches annually, and that areas with more than sixty inches are doomed to remain moorland. It is obvious that the majority of areas with more than fifty inches of annual precipitation are actually moorland, and in Scotland the moorland edge corresponds even more closely to the thirty-five-inch isohyet. The rate of evaporation, on the other hand, must be considerably less than twenty inches per year. An annual excess of moisture, which commonly more than equals the evaporation rate, produces serious drainage problems in all moorland areas, particularly in the exposed west and on the level uplands.

Poor drainage in moorland areas is aggravated by the absorbent humus of moisture-tolerant plants. Water-logged surface humus fosters the growth of such peat-forming plants as sphagnum, whose acidic remains accumulate as raw peat. Blanket peat, originating in this fashion, forms an intractable shroud over those British uplands that receive more than fifty inches of precipitation per annum. The blanket peat soils of the moorlands result solely from the accumulation of organic matter. The roots of plants growing on such peat are unable to penetrate to the mineral sub-stratum, and are therefore essentially divorced from it.

Types of Moorland Vegetation

Vegetation provides the most readily available, easily observable index to variations within the moorland environment. Moorland vegetation has changed considerably since prehistoric times, when forests are believed to have covered a large portion of the land below the 2,000-foot contour. Although the majority of these forests were cleared by man, his activities may have been merely an intensification of vegetation changes resulting from post-glacial climatic. Man has been an effective agent, however, in preventing the natural regeneration of forests, both by frequent fires and heavy grazing of his animals. The moorland cattle of the eighteenth and nineteenth centuries were largely replaced by wether [castrated male] sheep, whose selective grazing has virtually denuded moorland areas of their better forage plants, yet left the poorer to multiply untrammeled. The shift from wether to lamb production after the middle of the nineteenth century has accelerated the process of pasture deterioration. It is noteworthy, however, that changes in the botanical composition of moorland areas are largely reversible if sheep are removed.

Although the effects of man and his animals have wrought changes in moorland vegetation, these changes are now understood sufficiently well to warrant the use of the existing vegetation as an index to the total natural environment and to the improvements which are feasible under present technological and environmental conditions. We shall consider four major types of moorland vegetation in order of decreasing moisture and increasing drainage or desiccation. They are: (a) "moss," or bog, (b) Molinia-Nardus grassland, (c) heather moor, with an important sub-type, heather fell, and (d) bent-fescue grassland.

"Moss," or bog, is associated with the heavy precipitation and water-logged peat soils of "climatic moors." Water-loving plants, such as the sphagnum mosses (*Sphagnum spp.*), cotton grass (*Eriophorum vaginatum*), and deer sedge (*Scirpus cespitosus*), are dominant, although there are a number of other species. Peat is actively forming under these spongy, water-logged plants. Large areas of moss are too wet to be of value for either shepherd or forester without extensive drainage, but small mossy patches are valued by the shepherd because of their succulent early growth in spring, when other forage is to a large extent unavailable.

Drainage, either by man or natural erosion, will ordinarily convert moss into a grassland dominated either by Molinia (*Molinia cerulea*) or Nardus (*Nardus stricta*). Massive tufts of blue-green Molinia dominate level to undulating uplands with slowly percolating ground water, where inadequate drainage induces the formation of blanket peat, but is not quite poor enough to permit moss to become dominant. Nardus grassland develops particularly well on slopes below the level uplands and above the higher improved pastures. The acid soils of Nardus grasslands frequently are covered with peat by slope wash from the uplands, but Nardus is never associated with raw peat formation, so blanket peat is absent. Areas dominated by Molinia-Nardus grassland may be greatly improved for grazing purposes by plowing, fertilizing, and re-seeding them with better grasses, although Molinia areas often require some drainage as well. In addition, areas of Nardus are well suited to afforestation with Norway spruce (*Picea abies*), and Molinia land will produce Sitka spruce (*Picea sitchensis*) if the spruce is interplanted with nurse rows of Scots pine (*Pinus sylvestris*).

Heather moor, the third type of moorland vegetation, is commonly developed on the thin peat or podzolic soils of ancient woodlands. Common heather, or ling (*Calluna vulgaris*), is the dominant plant, but in small poorly drained areas one finds the heaths (*Erica spp.*), the moor rush (*Juncus squarrosus*), and even deer sedge, cotton grass, and sphagnum moss. Heather is intolerant of poor drainage; therefore, it is found only on steep slopes in areas of heavy rainfall, and on deep peat only when it is well drained. Young heather provides valuable grazing for sheep, but must be burned regularly every nine to fifteen years to prevent its becoming old and stemmy.

The bare rock, talus slopes, and boulders associated with glacial erosion support a mixed but sparse vegetation on the soil that collects in crevices or between rocks. Such an area is called a *heather fell*, after the dominant species, but any of a number of other plants may be dominant locally. Heather fell provides extensive grazing for hardy sheep, but, for physical reasons, it is virtually unimprovable. Afforestation of the lower fells is feasible only with the relatively unsalable Lodgepole pine (*Pinus contorta*), so taken by and large, the fell areas may be considered the least improvable land of Britain.

Bent-fescue grassland dominates the best moorland grazings and the poorest permanent pastures. Bent (*Agrostis spp.*) and fescue (*Festuca spp.*) grasses are found on well-drained slopes with shallow soils of moderate acidity; the grassland they dominate will deteriorate to poorer species unless carefully managed. Even under the best management, it is difficult to prevent some colonization by certain undesirable plants.

The most common and most pernicious of the subdominant species of the lower moorlands is bracken (*Pteridium aquilinum*), an aggressive fern which rapidly invades bent-fescue grassland, heather moor, or Nardus grassland. The shade of the bracken fronds stifles the growth of grasses and even of heather; therefore, it is quick to colonize and proves virtually ineradicable once established. Bracken is intolerant of poor or waterlogged soils, and shuns exposed sites, but rapidly takes possession of the better land which can least be spared. Although the shepherd realizes that bracken has taken his better land, the difficulty of eradication frequently prevents any effective reclamation on his part. The Forestry Commission is happy to acquire bracken-infested land, as it is well suited to the growth of Japanese larch (*Larix leptolepis*), but the shepherd is commonly reluctant to relinquish his best hill land, unless though it may actually be.

Gorse (*Ulex spp.*) is another common plant of the higher improved pastures and lower moorlands. Although its flower injects a glorious note of yellow into the drab early spring landscape, gorse is merely a pest to the shepherd. In addition to scattered clumps or bushes of gorse, the lower moorlands are also characterized by rushes (*Juncus spp.*). Rushes are common wherever the soil is moist or the groundwater table is high, and particularly along "flushes," where rivulets trickle down from the higher moors. Although rushes have some slight grazing value when the ground is covered with snow, the presence of gorse and bracken in permanent pastures is one of the first obvious signs of deterioration and abandonment.

The theoretical distribution of vegetation types, and the changes resulting from improvement or abandonment of the land, can be portrayed for a representative upland valley (Fig. 3.2). The valley bottom and gentle lower slopes are enclosed for crops or improved pasture. The higher enclosed fields, with slightly less gentle slopes, carry bent-fescue grassland, which might be improved to a richer pasture of rye-grass and wild white clover if properly fertilized and carefully managed; if neglected, on the other hand, the bent-fescue pasture will be invaded by bracken, gorse, and coarser moorland grasses, such as Nardus. The steep slopes above the head-dyke [stone wall at the upper end of a field] are dominated by heather moor, or "black moor." Heather must be burned periodically if it is to provide optimum grazing; careless burning, however, or excessive grazing create suitable conditions for the invasion of bracken or Nardus. The poor state of many upland grazings is indicated by the prevalence of "white moor" dominated by light-colored Nardus with scattered patches of green bracken and olive-drab heather. Above the valley slopes, the level uplands are characterized by "green moor" with Molinia dominant, where ground water is percolating slowly, or by moss if ground water is stagnant. Peaks rising above the level uplands commonly have heather moor on the thicker soils of their gentler slopes, but the steeper rocky slopes have only heather fell.

Agriculture in the Tweed Basin

The interdependence of upland and lowland farmers in Britain is nowhere better illustrated than in the basin of the Tweed River in southeastern Scotland and northeastern England, although it must be noted that sheep dominate the area to the virtual exclusion of cattle. The upper Tweed and its tributaries have cut deep, narrow "dales" into the surrounding hills, most of which rise well over 1,000 feet. The lower Tweed, however, flows in a broad rolling lowland less than six hundred feet above sea level. The higher hills receive heavy annual precipitation, which totals sixty to seventy inches on exposed summits in the west. The eastern lowlands are so shielded, however, that the majority of areas below eight hundred feet receive an annual average of less than forty inches of rain. Areas below four hundred feet commonly receive less than thirty inches.

The best land, lying in the lower Tweed valley, is devoted almost entirely to crop cultivation. Although most of the good quality land is also cultivated, it contains scattered fields of permanent grass because the annual precipitation is so heavy as to make cereal ripening uncertain. Land of medium quality is used almost entirely for permanent grassland. Most of the medium quality land lies on lower slopes of hills between 600 and 1,000 feet above sea level, but long tongues of both good and medium quality land penetrate poorer areas along stream valleys. This is particularly true of the dales in the western part of the basin. The hills above these valleys, and virtually all areas more than 1,000 feet above sea level, are of poor quality, and are used almost exclusively for rough grazing.

Hill sheep farms dominate the poor quality land of the Tweed basin. Virtually all of the land of the farm is moorland, although gentler slopes at lower elevations may be used as permanent improved pastures, and a bit of better quality land in the valley bottom may be cultivated or mowed for hay. The moorland grazings will keep a ewe alive over winter, and will sustain her and her lamb during the summer, but the hill sheep farmer is unable to fatten any stock or produce any appreciable amounts of winter feed. In the fall, therefore, the six-month old lambs and the over-age "cast" ewes are sold to farmers on better land for cross-breeding or fattening. A better understanding of the character of hill sheep and stock rearing farms will result from a consideration of an actual farm of each type in southern Scotland.

Auchencairn—A Hill Sheep Farm

Auchencairn farm is situated on the steep eastern slopes of Nithsdale in Dumfriesshire (Fig. 3.3). The eastern part of the farm is a rolling upland broken by knobby ice-scoured knolls. This upland ends abruptly in the western part of the farm, where a steep slope drops sharply down to a small floodplain. Crop production is inhibited by an annual precipitation which averages about fifty inches, so the upland is moor and the steep slopes are used almost entirely for permanent pasture. The moorland edge lies between 600 and 800 above sea level, and in general follows the crest of the steep western slope of the farm.

The upland is rather sharply divided into "black" heather moor in the south and "green" Molinia moor in the north. Heather is dominant, particularly on thin-soiled knolls, and patches of heather occur on such knolls in areas dominated by other plants. The dominance of Molinia in the north seems to be due to slightly deeper soil and poorer drainage, but the areas of poorest drainage are the boggy mosses between heather-covered knolls in the south. Molinia tends to give way to Nardus along the crest of the slope in the west, where drainage is better, but the two grasses are too intricately interdigitated to be shown separately. The wood lots along the upper slopes provided shelter and wood for the farm before World War II, but these were cut for timber during the war, and the farmer does not feel that he can afford to re-plant them. The formerly wooded areas are now dominated largely by bent-fescue grassland which is rapidly being invaded by bracken and heather.

The improved land of the farm is almost entirely grassland. The permanent grasslands, or "parks" as they are commonly known, are on the steep western slopes of the farm. The higher and more remote parkland is rapidly deteriorating to bracken. Some of the lower parkland was considerably improved when wartime subsidies were used to cultivate it, but crop production proved definitely uneconomic, and the land was returned to permanent grass as soon as the emergency ended. Portions of fields along the stream are liable to flooding, so they are used for pastures or hay meadows. The only cultivated land on the farm is the small fields of rape and oats near the farmstead.

Hill sheep farms are commonly divided into hirsels, a hirsel being the amount of land required to support the number of sheep (about 600) which can be tended by a single shepherd. It is desirable to maintain the same flock on the same hirsel; as the sheep acquire knowledge of the ground, they develop

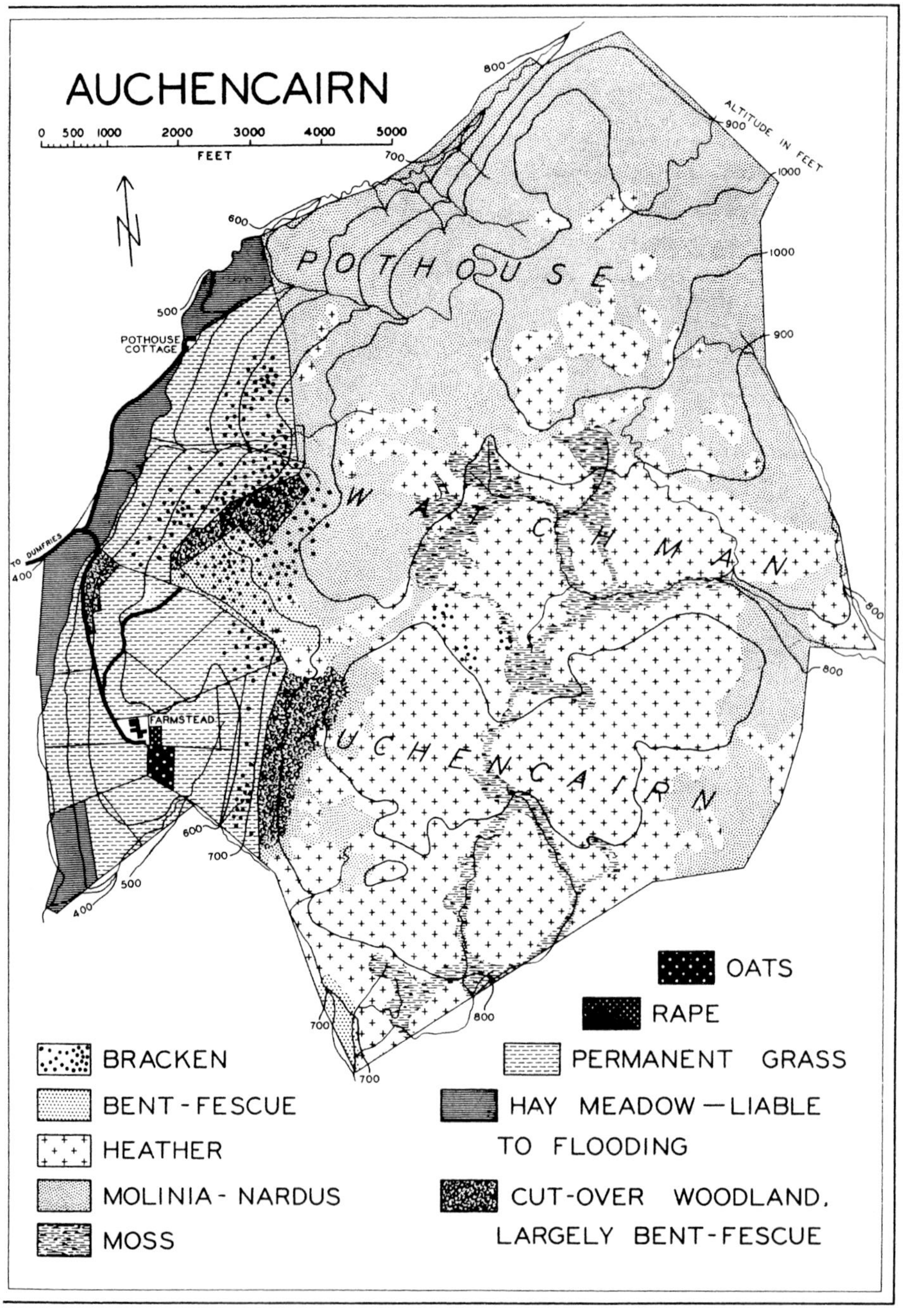

FIG. 3.3. Auchencairn—a hill sheep farm.

hardiness and resistance to disease, and they are available in regular ages for flock replacement. The introduction of sheep onto strange land is so difficult and liable to disaster that it is attempted only as a last resort. As a result, many hill sheep flocks are "bound" to the land; the incoming tenant must purchase the flock, with the agreement that the landlord will re-purchase it if the tenant leaves the farm.

Although Auchencairn is operated by the owner, the value of a self-sustaining flock is witnessed by the fact that Auchencairn flocks have been on the farm for more than forty years. The farm is divided into three hirsels. Pothouse and Watchman are tended by a shepherd who lives in Pothouse cottage, whereas the parkland and Auchencairn hirsel are tended by the farmer who lives at the farmstead. Although the farm is only ten miles from the county town of Dumfries, and has good bus connections, the farmer has difficulty keeping a shepherd. Pothouse hirsel, largely Molinia, is grazed by about 350 Cheviot ewes, whereas the heather of Watchman is grazed by about 190 Blackface ewes.

Spring is the season for lambing, shearing, and marking on the hill sheep farm. Lambing coincides with the first flush of spring vegetation toward the end of April. After lambing, the ewes and their lambs are allowed to graze, disturbed only by the daily round of the shepherd, until they are gathered for shearing in late June. The wool is delivered immediately after shearing to a wool merchant designated by the government, and the farmer is paid a fixed price for it. Lambs are not counted until shearing time, when wether lambs are castrated, and all lambs have their tails docked and are marked. The number of lambs marked per hundred ewes in the breeding flock the previous fall is known as the "percentage of lambs marked." Although this percentage is influenced by winter ewe losses, lambing losses, and spring losses of ewes and lambs, it is the farmer's only index to flock fertility. Watchman has a marking percentage of seventy-eight percent and Pothouse of seventy-three percent, neither particularly good.

The major problem of the hill sheep farm is the provision of winter feed, so all surplus stock must be disposed of in the fall. Both sheep and lambs are dipped in late July to rid them of ticks, lice, and other vermin before the fall sales. All of the ewe lambs are retained for flock replacement. Cheviot wether lambs are sold as stores, but some of the Blackface wether lambs are fattened on rape and sold for direct slaughter. As a general rule, Auchencairn annually sells one lamb for each two breeding ewes. After the lamb sales, the farmer prepares his flock for winter by casting ewes, selling surplus young ewes, and buying rams. Old ewes are cast at six years, after they have produced five lambs on the hill. If the winter has been mild, with small ewe losses, there may be more year-old ewes ("gimmers") than are needed for flock replacements, so these too are sold. In turn, the farmer must purchase some new rams each fall. The rams are turned onto the hill grazings in late November, and remain with the ewes until New Year's Day, but are kept on lower pastures for the rest of the year.

Winter is a time of semi-starvation for hill sheep. The ewe lambs kept as flock replacements are trucked to a lowland dairy farm for the winter months. The dairy farmer makes a profit by renting his pastures while cattle must be housed and the sheep farmer profits by getting his ewes off to a good start. From the time they return home the next spring, however, until they are cast, the ewes spend their entire lives on the hill. In spite of visual evidence to the contrary, it would seem impossible for an in-lamb ewe to maintain herself, to withstand the rigors of an upland winter, to supply the necessary nourishment for her developing lamb, and finally, to give birth to and successfully rear her offspring. It is remarkable that she is able to do this at all, and when she does, it is at the expense of her vital reserves; by spring, she can only be described as emaciated. The annual loss of Cheviot ewes on Pothouse runs around one-third, whereas three-quarters of the hardier Blackface ewes of Watchman survive the rigors of winter.

The farmer tends the 280 Cheviot ewes of Auchencairn hirsel himself. Only about 200 lambs are marked annually, and the ewe loss runs about thirty percent. These Cheviot ewes thrive on heather-covered hills, contrary to the popular belief that only Blackface sheep do well on heather. Strips of heather moor are burned in rotation every five years, as the plant becomes so stemmy after that time that it provides little forage. The hirsel would be improved if the mosses could be drained, yet they have considerable value in early spring, when they produce shoots of new growth long before either heather or Molinia.

The farmer also manages the sheep and cattle of the park. A flock of 135 Cheviot ewes is tupped by Border Leicester rams to produce heavier half-bred lambs. The lamb marking rate is ninety percent in the park, which compares favorably with seventy-eight on Watchman, seventy-three on Pothouse, and seventy-two percent on Auchencairn hirsels. The parkland also supports a herd of about 110 Galloway cattle, which produce store animals for sale at an age of a year and a half. About a quarter of these cattle are grazed on the hill from May until November, to use the hill grazing at their lushest, and at the same time to permit hay to be taken from the park. All of the hay and oats are required as winter feed, but the rape is surplus to farm requirements and is used to fatten Blackface lambs for slaughter.

Enzieholm—A Stock Rearing Farm

Enzieholm farm is situated on the south side of Eskdale in eastern Dumfriesshire, where the annual precipitation averages forty-five inches. From the floodplain of the River Esk, the farm slopes up to a height of 1,115 feet above sea level (Fig. 3.4). The eastern and western boundaries of the farm parallel two rushing burns [small streams] whose steep-sided valleys are overgrown with scrub woodland of little economic value. Scrub woodland also covers the steep banks of a stream terrace south of the road. The lower parts of the floodplain are frequently inundated and are used only for hay meadows, but crops are grown on the fertile soils of the more elevated parts. The gentler slopes south of the road are used for a six-course rotation of oats, roots, oats, grass, grass, grass, but the steeper fields are used as permanent pastures. The farmer plans to eradicate bracken, which has begun to invade his higher permanent pastures, by plowing, fertilizing, cropping, and then re-seeding them. A crop of oats is being taken from a field which had not been cultivated for ten years, and the adjacent field will receive the same treatment next year, when the oat field will be seeded with a grass mixture.

The moorland edge lies between 500 and 600 feet above sea level. Lower bent-fescue grasslands are seriously infested with bracken, and bracken has even begun to invade some areas dominated by Molinia. Drier knolls within the Molinia-dominated area are covered with bent-fescue grassland, whereas poorly drained depressions contain mosses. The same flock of Cheviot sheep has been maintained at Enzieholm for thirty-six years, but they are not bound, even though the farm is operated by a tenant on a one-year lease. A third of the ewes are bred pure to maintain the flock, and these ewes remain on the hill throughout the year; the other two-thirds are crossed with heavy Suffolk Down rams to produce a heavy store lamb. The ewes carrying Down lambs would not survive a winter on the hill, so they are brought down to the parkland before lambing in the spring. The annual ewe loss is only seven percent, but the percentage of lambs marked is only seventy-three. Pure Cheviot ewe lambs are saved for flock maintenance,

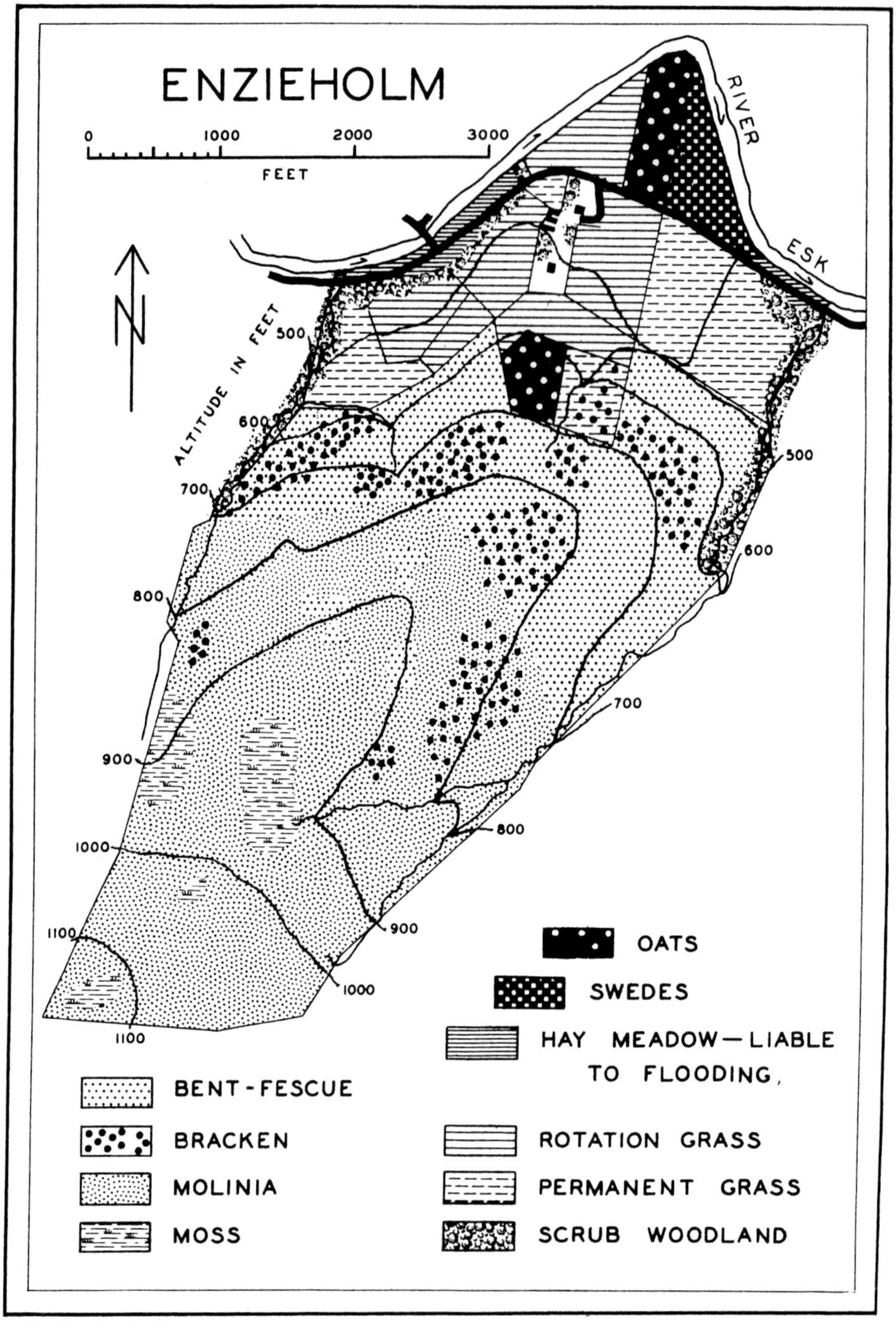

FIG. 3.4. Enzieholm—a stock rearing farm.

and Cheviot wether lambs are fattened on the farm for slaughter, but all of the heavy Down Cross lambs are sold as stores. Ewes are cast after their fifth lamb.

In addition to sheep, the parkland carries a breeding herd of thirty Ayrshire dairy cows. Milk is used entirely for calf feeding or farm use and none is sold. Calves are born in the spring, spend a summer on the hill, a winter in the park, a second summer on the hill, and then are sold as stores in the fall when they are a year and a half old. The additional labor requirements of the dairy herd necessitate the assistance of a hired laborer, who lives in a cottage up the hill from the farmhouse.

Although more than three-quarters of Enzieholm is moorland, and the total farm area is less than one-fifth that of Auchencairn, Enzieholm has three times Auchencairn's acreage of crops and rotation grass. The winter food produced from this greater acreage of arable land is used to diversify the production of the farm. The moorland of Enzieholm is used as supplementary pasture, and practically ignored, whereas the moorland of Auchencairn supports the hill sheep which are the major source of farm income. Significantly increased production from either farm will result primarily from increasing the production of its moorland because the improved land is already used intensively.

Moorlands for Forestry and for Hill Farming

The basic land use problem of the British moorlands is the equitable allocation of land to forestry and other uses without reducing food production. A large moorland acreage will continue to be dominated by hill sheep farming within the foreseeable future, but forestry will likely play an increasingly important role. Land for the five million acres of [new] woodland proposed by the Forestry Commission as a minimal goal must be obtained from the hill sheep farmers who now occupy it. The problem of obtaining land for forestry is complicated by the present policy of the Ministry of Agriculture in England and Wales and the Department of Agriculture for Scotland. These two governmental agencies are maintaining hill sheep farming in a state of artificial prosperity with the hill sheep subsidy. In addition, they commonly are able to prevent the use of any farm for purposes other than grazing by allowing it to be improved under the provisions of the Livestock Rearing Act of 1951.

Areas best adapted to hill sheep farming should be improved for grazing so that they will produce even more food, whereas areas of low productivity should be afforested. Ideally, the best land of each farm should be improved, and the poorer and more exposed land should be planted in wood lots and shelter belts. In practice, however, woodlots and shelter belts are not conducive to economic forest management; perhaps the less productive farm units may have to be afforested *in toto*. It is probable that all moorland which cannot be economically improved for agricultural use should eventually be afforested. The poorest and highest moorlands, especially those more than 1,500 feet above sea level, which at present do not appear afforestable, would seem to hold considerable potentialities for recreational development.

Prosperous hill sheep farms are essential to Britain both as direct producers of mutton and wool and as reservoirs of breeding stock for lowland farms. It is important to remember, on the other hand, that sheep are losing their importance in many lowland areas which they formerly dominated. Tavener (1952) shows that a general farming system producing cereals, cattle, and root crops has replaced the sheep-cereal-roots system which formerly dominated the chalklands of Dorset. It appears that cattle will continue to replace sheep in Lowland Britain. The hill sheep farmer will thus lose his market for store sheep; therefore, he must either fatten more of his own sheep or turn to the production of store cattle for lowland farms.

The more productive hill sheep farms should be improved with the goal, in the majority of instances, of doubling their present carrying capacity. The better hill farming land would ideally be organized into family-operated farms supporting about 500 sheep, twenty breeding cattle, pigs, and poultry. Hill sheep farmers should also be alert for new techniques which will permit still greater production; one farmer has discovered that a shepherd in a jeep can tend more than twice as many sheep as he formerly could handle on foot. Intensive improvement of the more productive farms would more than replace the food lost by the afforestation of less productive areas.

Land Division

CHAPTER 4

in Britain*

The arrangement of man-made structures on the landscape cannot be understood without some knowledge of the history of settlement, the division of the land, and the acquisition of title to it by the settlers. One of the necessary first steps in the human occupance of any part of the earth's surface is to decide who has what rights to use which parts of it. The way in which the land is divided has a strong influence on administrative boundaries, property boundaries, field boundaries, methods of cultivation, the arrangement of farmsteads, and the layout of roads.

A concern about the division of land, and rights to use it, appears to have developed in association with the development of agriculture. Primitive hunting, fishing, and collecting groups appear to have rather vague concepts of land division. The agriculturalist, however, feels that he has a right to the products of the land after he has tilled the soil, planted the seed, and tended the crop. He demands that his fellows respect his right to his plot of ground and what he can produce on it, but in return he is prepared to accord equal respect to their rights to their plots.

The right to use land may be held by an individual, or it may be held in common by all members of a tribe, community, or society. This right may be exercised by the owner, or it may be extended to another user, usually in return for some consideration, such as the payment of rent, whether in cash or in kind, or the performance of specified services. In order to minimize conflict between

---

* "Land Division in Britain," Chapter 3 in *The Look of the Land* (Englewood Cliffs, N.J.: Prentice Hall, 1975).

the owner and his neighbors, it is customary to demarcate the limits of any right to land by drawing a boundary around the land to which the right pertains. The overall system (or lack thereof) whereby such boundaries are drawn is a system of land division. Systems of land division fall into two broad categories: (1) those established before the land was occupied and settled have antecedent boundaries with regular geometric patterns; and (2) those established coincident with, or even subsequent to, settlement have irregular boundaries which follow easily distinguishable natural features, such as streams and ridge lines.

Early Land Division in Britain

The earliest recognized human inhabitants of Britain were Paleolithic bands who roamed the edge of the forest in small kin groups, the men hunting wild game, and the women and children collecting everything their experience had taught them was fit to eat. Their material possessions were scanty, and they huddled for shelter in caves along the coast or near the chalk uplands where they could find the precious flint they needed for their stone tools. Perhaps these groups had their own "hunting grounds," and recognized the territorial rights of others, but the Paleolithic people of Britain have left little trace upon the landscape.

The first enduring mark was made by the Celts, groups of farmers who arrived around 750 B.C. and began to build small isolated clusters of thatched huts and barns along the hill crests. The Celts used a light, bronze-tipped, ox-drawn plow to cultivate small square plots of half an acre or so, enclosed by banks of earth or stones. These plows dug a straight furrow and cut the sod, but they failed to turn it because they had no moldboards, and thus it was necessary to cross-plow the field at right angles to secure the desired tilth. Cross-plowing produced squarish fields which still show clearly on aerial photographs; these are commonly called "Celtic fields," although they have no relationship to the modern patterns in any Celtic lands. Most of the fields which can still be seen are concentrated on the open chalk uplands, which have not been cultivated in subsequent millennia, but traces of such fields in other parts of England may have been obliterated by the activities of later plowmen.

Shortly before the Roman invasion, another group, the Belgae, arrived in England with iron tools and heavy plows drawn by teams of eight oxen. Unlike previous invaders, who had been more or less restricted to the light soils and open country of the chalk uplands, the Belgae began to attack some of the lowland forests, because they had iron axes with which to fell the trees, and plows heavy enough to till the heavier lowland soils. They plowed long furrows, to avoid having to turn the cumbrous teams of oxen more often than necessary, and their long, narrow fields contrast sharply with the small squarish "Celtic" fields.

After the Roman invasion some of the native farmers prospered and adopted Roman ways of life. The Roman villas of England consisted of solidly built farm houses, barns, and outbuildings set around a courtyard amidst the large open fields of the farm. It is impossible to say how many were built by native farmers, but this was probably one of those times in English history—the sixteenth and seventeenth centuries were another—when prospering farmers erected proud mansions to advertise their new-found success and rising status.

The most significant Roman contribution to the rural landscape was the construction of more than 5,000 miles of roads, many running straight as an arrow across the countryside. The lines followed by many of these roads are still in use today, but of greater consequence is the fact that they provided ready-made routes whereby the later Anglo-Saxon invaders could penetrate wild and unsettled country more readily than if they had had to cut their own way through it.

The Celtic Fringe

As each successive wave of invaders from the continent descended upon the lowland plains of southern and eastern England, the unhappy native people fled farther into the hills to the north and west. These rugged uplands, swept by wind and lashed by rain, are hostile to man and beast. Their tough old Paleozoic rocks, stumps of ancient mountains, have been scraped bare by glacial ice or plastered by blankets of bog and acidic, infertile peaty soils; the small pockets of even moderately good soil are few and far between. This forbidding land became the refuge and home of the Celtic people.

The Celts evolved an infield-outfield agricultural system which was in marvelous harmony with the inhospitable environment; for the most part, it has disappeared in modern times, but it can still be found in a few remote areas. The infield was the best patch of ground available. It was enclosed by a stone wall or high earthen bank, received all of the manure, and remained in continuous cultivation. The standard crop was oats, which nourished both animals and people. The outfield, of lesser fertility, was essentially pasture land, but it was systematically cultivated under a system of "long ley." A patch of the outfield was enclosed by a temporary "fold" the summer before it was scheduled for cultivation, and cattle were penned on it to enrich it with their droppings. Then, after a crop of oats, or perhaps two, had been taken from it, the land was returned to grass until the time came to cultivate it once again. Each part of the outfield was cultivated in this fashion once every seven to ten years. Regular cultivation was necessary to renew the grazing quality of the pastures, which deteriorated fairly quickly under the excessively moist conditions of the Celtic lands. Even today, many farmers in Scotland still talk about the necessity of "taking the plow around the farm" in order to maintain the quality of their pastures.

In many areas the uphill margin of the outfield, which was the upper limit of cultivated land, was enclosed by a stone wall, or head-dyke. The improved grasslands below the head-dyke and the rough grazings of heather and coarser grasses on the moorlands above it were used as pastures for cattle and sheep. Some animals were taken to distant upland pastures during the summer cropping season, which, in Ireland, traditionally began on St. Patrick's Day and ended on Hallowe'en. The young folk, and sometimes entire families, left their winter homes and took their livestock and even their furniture to the summer pastures, where they lived in a hut known as a *booley* in Ireland, a *hafod* in Wales, a *shieling* in Scotland, and a *saeter* in Norway.

The most common type of winter home was a low, one-story, rectangular cottage which sheltered the members of the family in one end and their cattle in the other. It was twelve to fifteen feet wide, of varying length, and built of any locally available material—stone, timber, clay, wattle, even sod. The cottage was built on a slope at right angles to the contour. The top end, with the chimney, fireplace, and living quarters, was excavated into the hillside for warmth. The lower end housed the cattle, with the slope permitting refuse to drain away.

The question of how these cottages were distributed in a dispersed or nucleated fashion has been debated. Because the Celtic lands have experienced more than one cycle of settlement dispersion, nucleation, and dispersion again, it has been suggested that invaders subjugated the native people and forced them to live as vassals in clustered settlements, whereas the invaders were free to live in isolated farmsteads. Inheritance laws required that a man's property must be divided equally among all of his sons, which could have converted an isolated farmhouse inhabited by a father into a cluster of farmhouses inhabited by his sons or grandsons. Ireland once had many small, irregular clusters of farmhouses (known as *clachans*) in which nearly everyone had the same last name and, presumably, a common ancestor.

Open Fields and Villages

The landscape of villages and strip fields, which is so characteristic of the lowlands of eastern England, owes much of its character to the Anglo-Saxon invasions. The Anglo-Saxon invaders came from the marshy shores across the North Sea to a country that was essentially still a wilderness; within six centuries they transformed it into a settled agricultural land. Few modern English villages can be traced back before about A.D. 450 when the Anglo-Saxon invasions began, yet the names of almost all of them can be found in the pages of the Domesday Book of 1086, which was compiled twenty years after the Norman Conquest. The Anglo-Saxons divided the land into townships, which became the manors of feudal times, then the parishes of the church, and eventually evolved into the smallest civil divisions of contemporary England.

The Anglo-Saxons came in small groups, often consisting of kinspeople. They were well-armed and pugnacious and were constantly clashing with the natives and other groups. They cleared the forests and converted the land into large, open, unfenced fields around their villages. Many of the villages were given the name of the leader of the group, plus a suffix, such as –ing, -ham, -ton, -ingham, or –ington; for example, Laxton in Nottinghamshire was originally Lexington—the *tun*, or farm, of the people of Leaxa.

The land of the village commonly consisted of a nucleus of cropland on the better soils, with meadows for hay along the stream or in the wetter parts, and woodlands or wastelands of bog, heath, and moor in the poorer marginal areas; rights to cropland were associated with rights to the meadow, woodland, and waste. Common cultivation of the cropland was virtually necessitated by the fact that few men were wealthy enough to own a heavy plow and the eight oxen needed to pull it; one man might contribute the plow, another a team of oxen, and so on. The standard unit of cultivation was the *strip*, or the amount of land that could be plowed in one day. The strip was a "furrow long" (furlong), or the distance that a team of oxen could pull the plow without having to stop for rest.

The ideal strip was about 220 yards long and twenty-two yards wide, or an acre in size.

A bundle of strips made up a more or less rectangular block known as a furlong; each furlong block had its own name, and may well have been the land that was cleared by the communal group in a single year. The largest unit of all, the field, was a crazy quilt of many furlong blocks of all sizes and shapes, containing hundreds of strips. The cropland of each village was commonly divided into three large open fields, each covering a few hundred acres, and containing many furlong blocks, which were cultivated in the standard three-year rotation of food, feed, and fallow.

The strips which each man farmed were scattered far and wide over the cropland of the village. They were apportioned so that he had an approximately equal number in each of the three fields, and thus the only sensible, central place for him to live was in the village. Neither the strips nor the furlongs, and quite often not even the fields, were enclosed by fences of any kind, thus creating the historic landscape of farm villages amidst open fields which was common throughout medieval Europe. By night the cattle and sheep were kept in their stalls, or in the fields in temporary enclosures known as folds; during the day they grazed the common wastelands, the fallow field, the meadows (after they had been mown for hay), and the barley stubble (after the cropland had been harvested in the fall). The lack of fences necessitated oxherds, cowherds, and shepherds to keep the sheep out of the meadow and the cows out of the corn. Swine, which were only semi-domesticated, ran loose in the woods in care of a swineherd. It is not surprising that, in the medieval ballads, when a traveler encountered a wild man in the forest, he was almost invariably there tending swine.

The land of each village was originally separated from the land of other villages by empty wastelands of forest, bog, heath, and moorland, and thus it required no sharp delineation. As population increased, however, these wastelands had to be reclaimed and brought under cultivation to meet the growing demands for land, and it became ever more important to demarcate sharply the boundaries of the village as the territorial interests and designs of neighboring villages began to come into conflict. Young men were expected to learn the boundaries, or *marches*, of their village, which were shown to them in an annual perambulation led by one of the village elders; the ceremony of "riding the marches" is still celebrated each year at Hawick in Scotland and perhaps elsewhere as well.

The Normans, Feudalism, and the Manor

Anglo-Saxon society recognized three classes of men: those of noble blood, or thanes; common free men, or churls; and slaves. The free men of the village maintained their lord by their labor, and in return expected leadership, protection, and justice. The average man had a lord of some sort to whom he gave presents, for whom he did jobs, such as harvesting and woodcutting, and under whose leadership he fought. In addition to working on his own land, each man worked a certain number of days on the *demesne* (his lord's) and the *glebe* (church) land.

The Anglo-Saxons combined communal cultivation with an acceptance of overlordship and the obligation of free men to cultivate the land of their lord. After the Norman Conquest, William the Conqueror replaced native overlords with foreigners, his own men, and installed the feudal system firmly and systematically in every corner of his realm. The feudal system William installed was the standard form of political organization in medieval Europe at a time when central government was too weak to maintain effective law and order, and the economy was almost purely agricultural, based upon barter and exchange of services rather than on money.

No one actually owned land outright under the feudal system; he merely held it as a hereditary right from some overlord in return for the personal services he rendered to his overlord. In simplest theory, the king, as God's chief servant in his kingdom by divine right of birth, held all of the land of the kingdom, and parceled it out among his nobles in return for the services, mainly military. He assigned principalities to princes, duchies to dukes, marches to marquesses, earldoms to earls, counties to counts, and baronies to barons; each of these, in turn, might divide the land into smaller units, which were assigned, once again, in return for military services. At the very bottom of the list was the smallest unit of land, the manor, which was just large enough to provide one gentleman or knight (the lord of the manor) with a livelihood for himself and his family, plus the horse, armor, and weapons he needed to render military services to his lord and to protect the peasants who worked on his land.

Each manor was an independent unit, and no two were exactly alike, yet they all had much in common. In its purest form, the manor consisted of a village in the midst of unfenced open fields, and all the land of the village (Fig. 4.1). Its most impressive structure was the home of the lord of the manor,

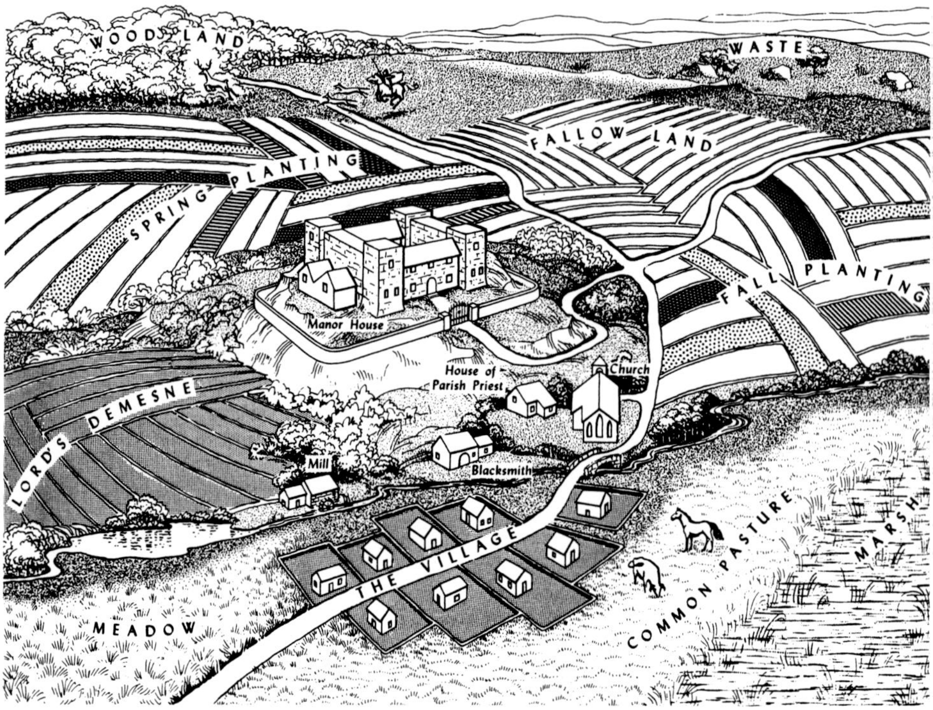

FIG. 4.1. Stylized diagram of a medieval manor. Reproduced by permission of the publisher from Wallace K. Ferguson and Geoffrey Bruun, A *Survey of European Civilization*, 4th ed. (Boston: Houghton Mifflin, 1969), p. 155.

or squire, which might have been a fortified castle in the early days, although later many castles were replaced by more comfortable mansions, or "halls." Close by the hall was the parish church and the home of the parish priest, for in many cases the church adopted the land of the manor as an ecclesiastical parish.

The main body of the village was the cluster of dwellings in which the peasants lived, and they all lived in the village. The village had no stores, but there was a smithy where the blacksmith made nails, shoes for horses and oxen, and plow irons. Beside the stream was a water mill, where the villagers were supposed to have their corn ground. The lord of the manor claimed the right to make all of his tenants grind their grain at his mill, and required them to give him a certain portion of it for this service. This claim was a constant source of friction, because the landlord often leased these rights to a miller for a fixed rent, and millers tended to take more than their fair share. Many of the villagers preferred to use the old-fashioned hand mills in which women were accustomed to grinding grain at home.

Although traces of the feudal past still linger on in such modern names as the Principality of Wales, the Duchies of Cornwall and Lancaster, the "Marches" of Wales and Scotland, and the ubiquitous "county" (which crossed the Atlantic Ocean to all fifty states except Louisiana and Alaska), the principal relic of feudal political organization is the basic unit of land-holding and decision-making, the manor. Over the centuries, in a very fuzzy fashion, the manor was largely superseded by the ecclesiastical parish for administrative purposes, perhaps in part because the poor laws were administered by the parish and in part because the relationship between manor, parish, and village was seldom as neat and tidy as the simplistic picture I have painted here. Nevertheless, village, parish, and manor were held together by the fact that the people of the village were expected to go, and for centuries did go, to the parish church, and the greater part of the village and the land of the parish probably was owned by a single squire, who was a key figure in decision-making as far as the use of the land was concerned. One might add, for the sake of the agricultural geographer, that the parish is the smallest unit of land for which agricultural statistics are available in Britain.

Enclosure and the Agricultural Revolution

The medieval manor produced all it needed, but needed all it produced; the manorial system was essentially a closed, domestic, noncommercial economy. There was little surplus for trade, and the exchange of goods consisted largely of barter at the local level. Landownership conferred prestige, but it had little monetary value at a time when the rendering of services, rather than money, was the principal medium of exchange. As the economy gradually became more commercial, however, and as money began to circulate more freely, both landlords and their tenants were increasingly tempted to "commute" customary services into payments of money; the prosperous tenant might prefer to pay rent instead of having to work on his lord's land, and the landlord might be quite happy to convert customary services into cash income. The desires of landlord and tenant quite often failed to coincide, of course, and the long history of the demise of feudalism is the sorry tale of a series of wrangles between landlords who insisted upon customary services, or demanded money instead of services, and tenants who insisted upon commuting these services, or demanded their right to perform them instead of paying rent.

The two most dramatic episodes in the decline of feudalism and the open-field system in England both occurred in times of rural depopulation and rapid agricultural change, and both were periods of fairly widespread enclosure of open fields by hedges, stone walls, and other kinds of fences. The first wave of enclosures is associated with the Black Death, which first broke out in England in 1348 and wiped out a quarter to a third of the population. The vastly reduced rural labor force was inadequate to maintain the traditional open-field cropping system just at a time when landlords were beginning to realize the profits which could be made from converting the open fields to pasture for sheep and cattle. In many villages, the landlord evicted the dwindling band of surviving villagers, demolished their mud-walled hovels, and enclosed the land for grazing.

The first wave of enclosures excited fierce antagonisms against the enclosing landlords and great public sympathy for the evicted villagers, and some rather ineffectual laws were passed which forbade enclosure. Public policy changed, however, around 1660 in response to the innovations which triggered the Agricultural Revolution, and the second wave of enclosures began to gather momentum. It reached a peak between 1750 and 1850 when private Acts of Parliament approved the enclosure of some 4.5 million acres of land in nearly 3,000 parishes, mainly on the Midland Plain. This second wave of enclosure coincided with the beginnings of the Industrial Revolution, for which it provided both food and labor. Many farm workers who were pushed off the land by enclosure were pulled into the burgeoning urban industrial centers.

The second wave of enclosures was pushed forward vigorously by progressive landlords, whose ability to innovate was severely limited by the restrictions of the open-field system. In the open fields, for example, it was impossible to grow a crop such as clover, which would still be in the ground after the traditional harvest date, because the villagers forcefully defended their right of common grazing on the stubbles after that date.

The Agricultural Revolution of the eighteenth century was an upward spiral involving the enclosure of the open fields, the introduction of new crops, a new rotation system, and larger and better livestock. The new rotation of wheat, oats, barley, and clover produced more and better feed, which enabled the farmer to keep more and better livestock. On his newly enclosed pastures, the farmer could improve his stock by controlling breeding practices, which would have been impossible on the old open fields. The larger, more numerous, and better fed animals produced larger quantities of manure, which was reflected in higher yields of grain when it was returned to the fields. Some crops were eaten directly in the fields by flocks of sheep which were penned inside temporary fences, or folds; this not only fertilized the soil and helped to maintain its fertility, but it also cut down on harvesting costs.

Not all of the land which was enclosed was used for crops, of course. Although many of the former strips of ridge and furrow have long since disappeared under the plow, in some areas enclosure removed the plow as well as the plowman, and, since enclosure, the land has been used mainly to pasture cattle and sheep. Over large parts of the English Midlands, for example, the old pattern of ridge and furrow is still clearly visible, especially when the sun hangs low in the sky and the shadows lie long across the ground.

The New Rural Landscape

The enclosure movement produced a new rural landscape, a patchwork quilt of small fields stitched together by hedgerows. The newly enclosed field was surrounded by a ditch, which marked its boundaries and facilitated drainage. Soil from the ditch was thrown up into a mound on the inner side, and this mound was planted with cuttings of hawthorn or other plants which would grow to produce a quickset (live) hedge. A good quickset hedge can last forever, because it can be *laid* or *pleached* to renew its stocktightness when it begins to grow old and thin. Each woody stem is half severed near its base, bent over into a horizontal position, and then woven around 4½ foot staves which have been driven firmly into the ground; the tops of the staves are woven together with twisted sprays of blackberry or willow. The stems quickly send up new shoots, put down adventitious roots, and soon the hedge is as good as it ever was.

On many manors, the first land to be enclosed was the demesne, especially if the lord of the manor had been able to make judicious trades and thus consolidate his open-field strips into a contiguous block of land; on many manors this was done long before enclosure. Today, the land of the old demesne is often called the "home far" of the village, parish, or manor, and its operation is managed directly by the landlord or his representative.

Awkward problems of farm layout were created when the land of a manor was enclosed and divided into individual farm units consisting of sold blocks of land, because all of the farm buildings were in the village. Quite often, therefore, the new farms were shaped rather like a slice of pie, with their points in the village where the necessary buildings were located (Fig. 4.2). A pie-shaped farm is quite difficult to operate, however, and eventually this difficulty was alleviated, at least in part, either by building a new "field barn" at the far end of the farm, or by building a completely new farmstead near the center of the farm when the older farmstead in the village began to deteriorate.

The village remains the home of the landless farm laborer. Unlike North America, where land is cheap and labor is dear, in Europe land is so precious that farmers are prodigal of labor, which is relatively cheap, and six or eight men might be employed on a farm which could support no more than a father and son in the Middle West. In the United States, the first adult male one meets on a farm is probably the farmer himself, but in England, where half a dozen men may be working in a field, the farmer is the man who still has his jacket on; the workers have taken theirs off and hung them on the hedge. The farm workers who live in the village form the lowest and largest group in a three-tiered rural social structure. Well above them is their employer, the farmer who manages the land, but the farmer himself is likely a tenant of the man who owns it.

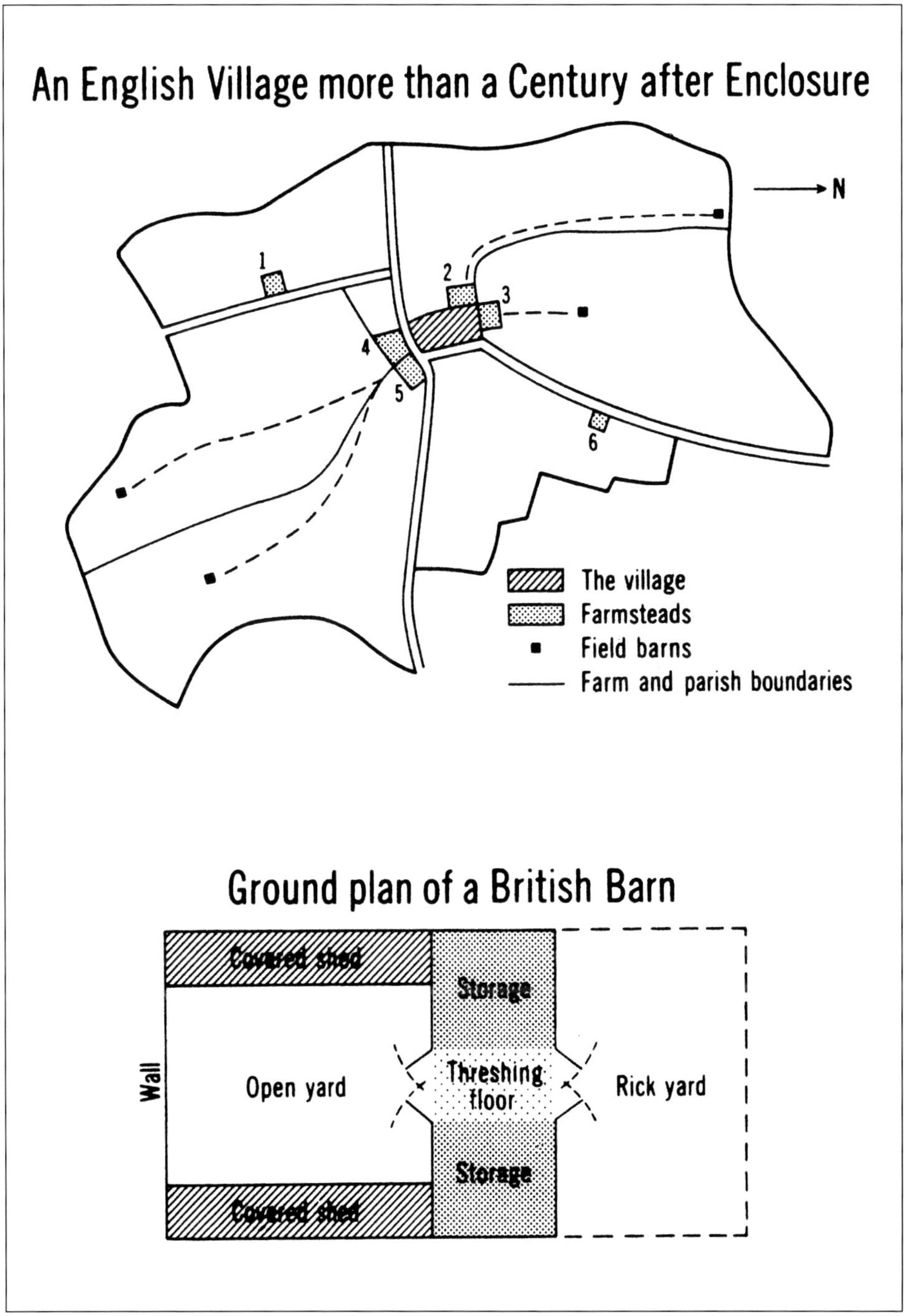

FIG. 4.2. The layout of farms in an English village, and the ground plan of a field barn.

Field Patterns

CHAPTER 5

in Indiana*

Unlike their European counterparts, American geographers have shown remarkably little interest in the study of field patterns. The lack of interest may be partly due to the fact that field shapes are extremely difficult to examine on the ground, and the United States Geological Survey did not begin showing field boundaries on its 1:24,000 topographic maps until 1957. In aerial photographs, however, field boundaries are clearly visible, and Marschner's classic work, "Land Use and Its Patterns in the United States," illustrates the astonishing geographical variety of field patterns in this country (Marschner, 1959).

An increasing familiarity with different parts of the state of Indiana led to the suspicion that field patterns can exhibit considerable geographical variation even within an area which has only one system of land division. Indiana is an appropriate place to test this idea because Indiana was the first state in which all the divisions of public land (except for a thin sliver along the Ohio line) were controlled by one principal meridian and one base-line parallel. The land-division system, with its rectangular coordinated township identification and number of sections, reached its final stage of development in Indiana and no further modifications to the federal surveys were introduced thereafter.

Four kinds of private land claims existed in Indiana before the congressional land surveys were made, and these claims were honored when the congressional surveys were carried out. Two originated in the deed of cession (1784) whereby the General Assembly of Virginia ceded to the United States all claim to the

* "Field Patterns in Indiana," *Geographical Review* 58 (1968): 450–71.

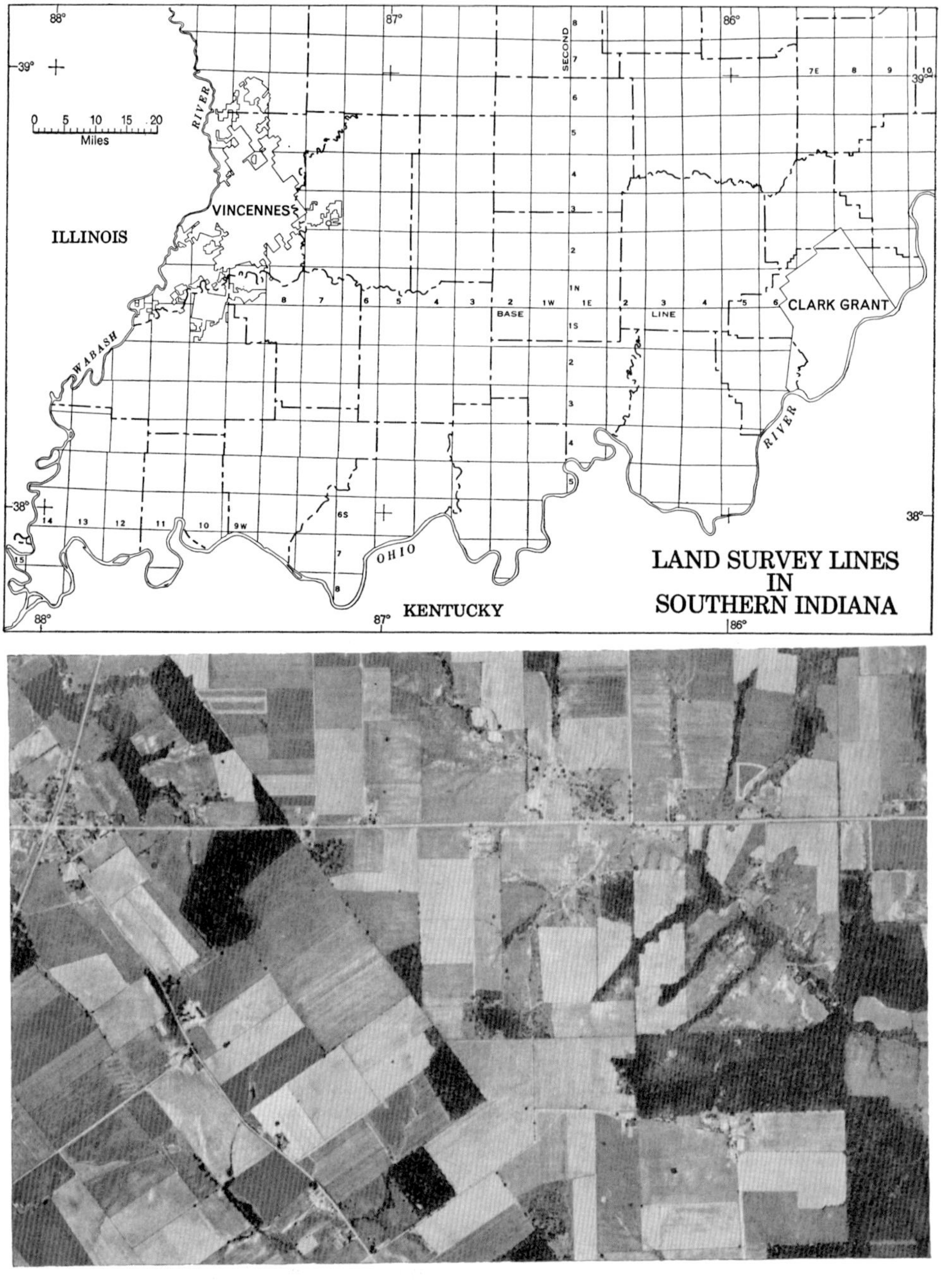

FIG. 5.1. (Top) Location of the Clark Grant and Vincennes—areas of noncongressional land survey—in southern Indiana. (Bottom) Aerial photograph southeast of the village of Nabb, Indiana, showing the boundary between the Clark Grant (lower left) and congressional survey lands.

territory northwest of the Ohio River. This deed of cession incorporated two important stipulations: first, "that the French and Canadian inhabitants, and other settlers of the Kaskaskias, Post Vincennes, and the neighboring villages shall have their possessions and titles confirmed to them"; and, second, "that a quantity of land not to exceed one hundred and fifty thousand acres . . . shall be allowed and granted to . . . General George Rogers Clark, and to the officers and soldiers of his regiment, who marched with him when the posts of Kaskaskia and Vincennes were reduced" (Dillon, 1859).

Clark and his officers selected a tract of land on the north bank of the Ohio River just above the falls, across the river from Louisville, Kentucky. The tract was laid out with a northwest-southeast orientation, more or less at right angles to the general course of the river. The boundary of the grant stands out clearly where it abuts on areas of the congressional land survey (Fig. 5.1).

Vincennes Area

The area around Vincennes is more complicated because it contains no fewer than five distinctive kinds of noncongressional land division: long lots, commons, donations, militia donations, and locations (Fig. 5.2). The French who settled the area were fur traders, not farmers, and they had little interest in, or concern with, landownership. Questions of title to land in the Vincennes area were first raised when Clark's men, who had seen the area and liked it, returned after the Revolutionary War with a keen interest in acquiring land. When the French titles were finally unraveled, most of the residents' land was found to consist of narrow strips "two arpents wide by the usual depth" (forty arpents) running back from the Wabash River. The long lots, which are too small to be shown individually on a small-scale map, were in three clusters on the former prairie land near the village: one upstream, a second downstream, and a third due south of the western part of the second. The fields were not fenced, and all the cattle were kept south of the village on a large fenced common that contained more than 2,500 acres of land.

In addition to their claims to the long lots and the commons, which were allowed, the residents of Vincennes had considerable claims for supplies they had furnished to the Virginia troops. These claims were settled by gifts of land, or "donations," in a large tract immediately east of the village to all heads of households who had been residents in 1783. The original square tract of 64,000

FIG. 5.2. (Top) Land survey lines in the Vincennes area. (Bottom) Villiage of Oaktown, Indiana, at right, showing contrast between Vicennes donation lands (lower left) and congressional survey lands.

acres, which was divided into 160 lots containing 400 acres each, was not large enough, and eighty-six additional donation lots, a total of 34,400 acres, were laid out around it. The donation lots are oriented in a northwest-southeast direction, in sharp contrast with the north-south orientation of the congressional land survey.

Heads of households in the Vincennes area who did not receive regular donations were given "militia donations" if they were enrolled in the militia in 1790 or had performed militia service. The militia donations, which consisted of only 100 acres each, were laid off in a block south of the White River, near the modern villages of Hazleton and Patoka. Other private claims were generally disregarded when the donations and militia donations were laid off, and certain individuals were thereby forced to forfeit legitimate claims. These people were given the privilege of "locating" new claims of up to 400 acres elsewhere. The new claims, which were known as "locations," are irregular in size and shape and are scattered throughout the area. Today, the Vincennes area has little field evidence of the old long lots; many have been consolidated into single farms, and they have been cross-fenced. Apart from the nuisance of owning land that straddles a township line and hence is taxed in two parts, the land-division system appears to create no problems for local farmers.

Types of Farming Areas

With the exception of the Clark grant, the Vincennes tract, and certain areas where land was granted for Indian reservations or for road or canal construction, Indiana is divided into six-mile square townships and square-mile sections in accordance with the congressional system of land survey. Within this system, the principal variations in the field patterns of the unglaciated southern third of the state, which has extensive areas of high relief, may be attributed largely to topography. Field patterns in the northern quarter of Indiana have also been influenced by topography, the choppy topography of glacial moraines. The central part of the state, however, an area known as the Tipton Till Plain, has level to only gently rolling terrain. Poorly developed moraines, many of which have less than thirty feet of local relief, break the general monotonous flatness only slightly. Many divides are poorly drained; streams are sluggish and seldom reach the rock floor except near the Wabash River.

The Tipton Till Plain was selected as a test area for the idea that variations in field patterns are related to variations in types of farming, because here the variations in topography and in systems of land division are so negligible that they may be considered constants. The Tipton Till Plain contains parts of three types of farming areas: a cash-grain farming area in the west, a grain-and-livestock farming area in the center, and a general farming area in the east (Fig. 5.3). In each of these three areas, a sample county was selected that was completely within the type of farming area and was subject to almost minimal urban influence.

Benton County, in the west, is in the cash-grain farming area. The farms are large and prosperous and highly mechanized, have comparatively few livestock, and concentrate on the production of soybeans and corn, which are sold at the local grain elevator. This type of farming is familiarly known as "B. C. M. farming": B for soybeans, C for corn, and M for Miami, because many of the farmers take off for Florida when they have finished their fall plowing and do not return until it is time to start farm operations in the spring.

Tipton County, near the center of the state, is in the central grain-and-livestock farming area. Farms in Tipton County are intermediate between those in Benton and Jay in size, prosperity, and mechanization. This is corn-hog country. Corn is the principal grain crop, and most of it is fed to hogs, which are the leading kind of livestock. In 1964, Tipton County had more than four times as many hogs per acre as Benton County, and more than three times as many as Jay, but fewer cattle per acre than either of the other two. The soybean acreage has doubled since World War II and, together with corn, accounts for more than half the total cropland in the county.

Jay County, in the northeastern general farming area, is slightly more rolling than Benton and Tipton and has twice as much land in pasture. Dairy cattle and poultry are more important than in the other two counties, and beef cattle are about the same. Farms in Jay County are undersized. It is generally accepted in the Corn Belt that a man is not a real farmer unless he has cash sales of $10,000 or more, and only a fifth of the farmers in Jay County fell into this category in 1964.

Statistical Analysis

A random sample of twelve aerial photographs was drawn for each county, and the field boundaries were traced from an area two miles square near the

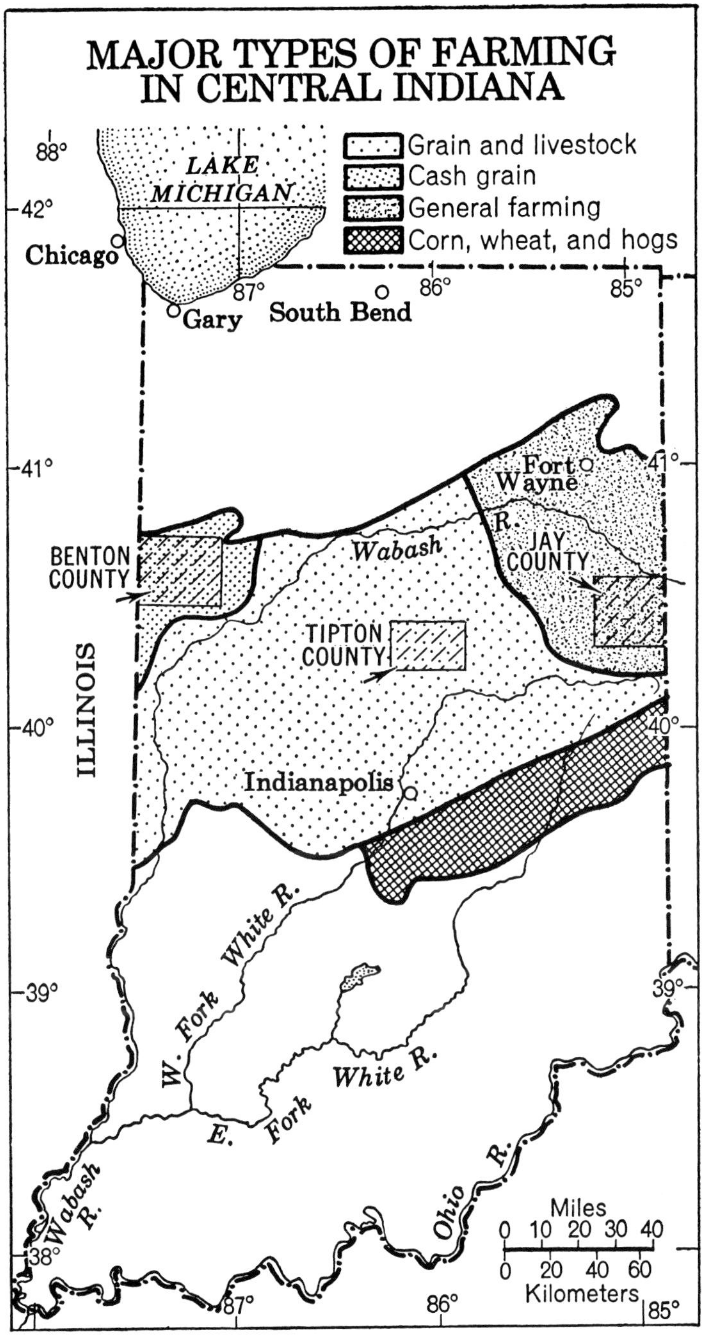

FIG. 5.3. Types of farming in central Indiana.

center of each photograph. The total area mapped in each county consisted of twelve plots of four square miles each, for a county total of forty-eight square miles, or 30,720 acres. A change in tone or texture in the photographs was taken as an indication of a boundary between fields. Streams and drainage ditches were not considered field boundaries unless they coincided with such a change. Farmsteads were treated as part of the field in which they were located.

Fields were analyzed with respect to size and shape, but orientation, which is so important a variable in areas of noncongressional land survey, did not appear worth measuring since virtually all fields were oriented either north and south or east and west. Field size was measured with a dot planimeter. Each field was classified into one of four shape categories according to the ratio of width to length. If the ratio was 1:1, the field was *square*; if greater than 1:1 and 1:2 or less, *rectangular*; if greater than 1.2, *elongated*; and, if the field was not rectangular, it was classed as having an *irregular* shape.

The average field size in Benton County, 26.9 acres, was two-thirds as large as the average field size in Tipton and Jay Counties (16.3 and 16.2 acres). When tested by analysis of variance, the difference in field size was statistically significant at the .001 level. The difference, however, is due solely to the difference between Benton County and the other two. The three counties had almost identical percentages of square and irregular fields, but Benton County had a far higher percentage of elongated fields (48.2 percent) and a lower percentage of rectangular fields (17.7 percent), whereas Jay County was just the opposite (34.8 percent and 30.6 percent, respectively).

In short, statistical analysis indicates that Benton County, which specializes in cash-grain farming, has fields that are both larger and longer than fields in the other two counties, and Jay County, with diversified farming, has a significantly larger number of rectangular fields. Tipton County seems to strike the happy medium as far as field shape is concerned. But why should the size and shape of fields in the cash-grain farming area be different?

The Field Check

One of the two-mile square sample areas in each county was selected for ground checking, which consisted of mapping the location and nature of all field boundaries. The results of the check were gratifying because they indicated that a change in tone or texture in the aerial photograph is a reliable indicator of a boundary between fields. Furthermore, the ground investigation indicated that farmers, like their fellow human beings, are creatures of habit, and the layout of their fields does not change much from year to year. The principal change that takes place is the enlargement of fields by the removal of fences and the joining of two small fields into a larger one.

Field enlargement is associated with farm enlargement, which has been going on in all three counties. The percentage of farmland in farms larger than 220 acres rose from sixty-four in 1940 to seventy-eight in 1959 in Benton County, and in Jay County the percentage rose from only sixteen in 1940 to forty-two in 1959. Farm enlargement has been associated with farmstead abandonment. Comparison with aerial photographs of the sample areas taken in 1950 showed that one farmstead in Benton County and three in Tipton County had been demolished; an additional seven farmsteads in Tipton County stood vacant in 1967.

Roughly three-quarters of all field boundaries in the three sample areas had the standard Corn Belt fence of woven (hog) wire, topped by a strand or two of barbed wire, on wood or metal posts. Barbed-wire and electric fences were found only in Benton County, where they made up less than ten percent of the total length of field boundaries. Many fences are obviously relict features, because the fields that they bound are not completely enclosed by fences, or because they are in sad disrepair. In general, fences are most common along railroad and highway rights-of-way, around farmsteads, and along property lines. The line fence is often the last to go, even when it is not maintained in stockproof condition, because "good fences make good neighbors," and a line fence obviates arguments about property lines.

The pattern of fencing differs considerably from one sample area to another, and the differences are closely related to differences in field patterns and types of farming. For instance, the sample area in Tipton County has the greatest length of field boundaries, the lowest percentage of unfenced boundaries, the highest percentage of internal fencing, and the largest proportion of square fields (Fig. 5.4). This is not surprising, because efficient livestock handling demands small and securely fenced fields. Most of the farms in the sample area still follow the standard corn-hog economy characteristic of the smaller farms of the eastern Corn Belt, and unfenced field boundaries are concentrated on a few large farms.

In the old horse-farming days, the field patterns in Jay County were much like those of contemporary Tipton County. The normal farm was eighty to

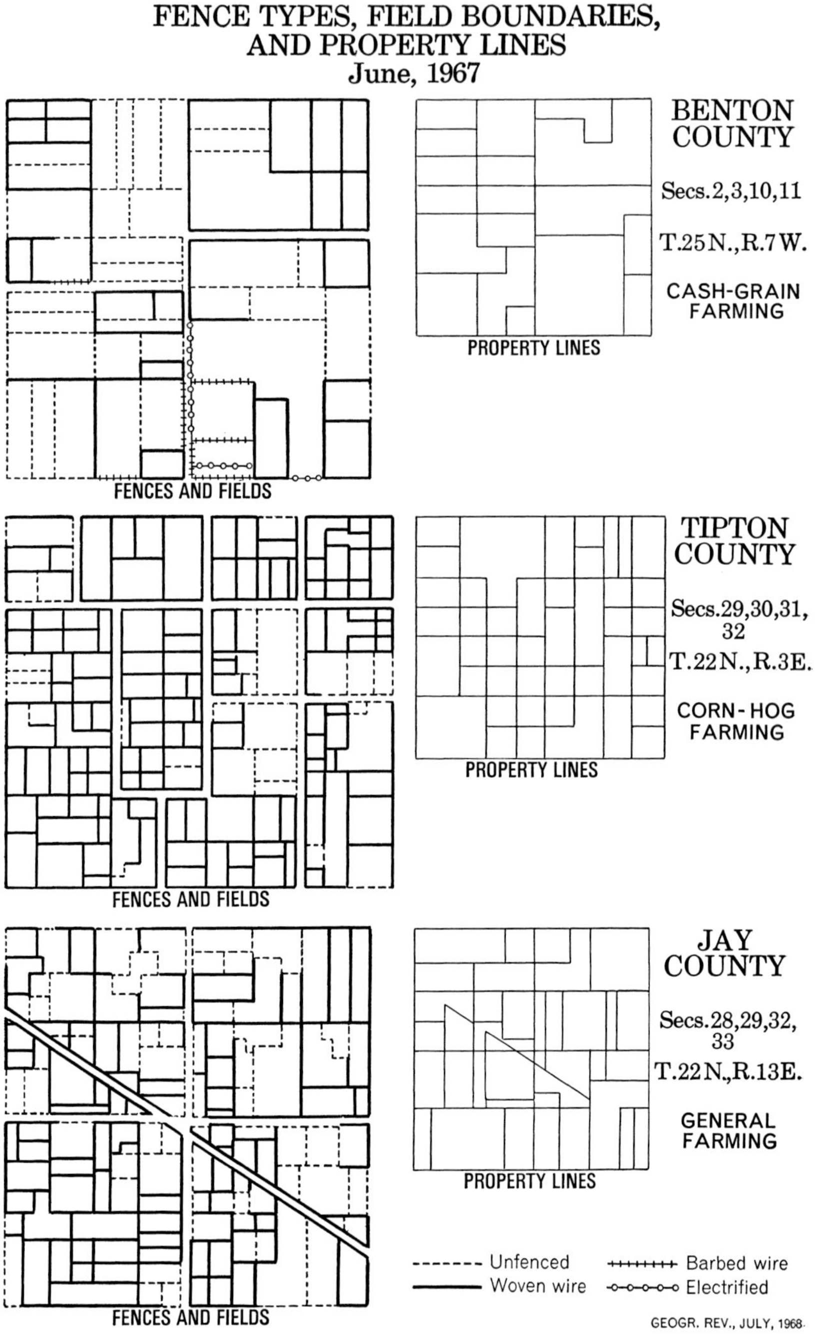

FIG. 5.4. Fence types, field boundaries, and property lines.

100 acres and was divided into four small fields, which fitted nicely into a four-year rotation of corn, small grain, leguminous hay, and pasture. The farmstead was located on a center lane with easy access to all fields, and the crops were fed to dairy cattle or hogs. But a 100-acre farm is too small to make efficient use of modern machinery and equipment, and local farmers have been forced to enlarge their operations, either by intensification or by expansion. Some have chosen to intensify by making the improvements necessary to qualify as producers of Grade A milk or by installing the equipment needed for confinement feeding of hogs. Although such operations can be conducted on a relatively small acreage, many of these farmers have to buy most of their feed, since they can produce only a fraction of what they need.

Other farmers have expanded their operations by buying more land or by renting land from former farmers who have taken off-farm jobs but are still living in the old farmhouses. Some of these farmers who have expanded have become beef-cattle feeders, others have switched to cash-grain farming. A successful beef-cattle operation in this area requires about 300 acres, whereas a cash-grain farm requires 500 to 600 acres. As the farm grows larger, and more of the farmer's capital is tied up in machinery, most human effort goes into producing cash-grain crops and there is little time left for livestock. Larger machines require larger fields with longer rows for greatest efficiency, and many small fields have been joined together by removal of the fences that separated them.

Fences have always been a nuisance, because the weeds and shrubs that grow up along them compete with crops for the plant nutrients in the soil. If the fence line is not cleared out each year it quickly grows up in trees, which can hurt crops as much as fifty to seventy-five feet away on each side, especially in dry weather. Even the farmer who keeps livestock has often no need for fencing because the animals are kept in a holding pen near the farmstead.

The removal of fences has gone farthest in Benton County, the cash-grain farming area. Benton County traditionally has had large farms and large fields, but fields are growing ever larger in the need to make the most efficient use of machinery. Less than half of the internal field boundaries in the sample area of Benton County are fenced. Although fences still stand along most of the highways and property lines, the majority are decrepit, and they will not be replaced. Fence removal is the order of the day in Benton County. Cash-grain farmers wish to run their rows the longest possible distance so that they will not have to waste time in turning machinery, and most farmers have rows that run the entire length of the farm; rows a mile long are not uncommon. This fact accounts for the large fields in Benton County, and also for the fact that so high a percentage of the fields are elongated.

To summarize, the field check substantiates the statistical conclusion that geographical variations in field patterns are related to geographical variations in types of farming. Tipton County comes closest to the traditional checkerboard pattern, because most of the farms are still practicing the traditional corn-hog economy. Field patterns in Jay County reflect the fact that the county's agricultural economy is in transition from general, diversified farming on undersized farms to more intensive farming on small farms or cash-grain farming on large farms. The large, elongated fields of Benton County have been created to enable cash-grain farmers to make the most efficient use of ever-larger farm machinery.

Tobacco Barns and Their Role in the

CHAPTER 6

Tobacco Economy of the United States*

Tobacco barns are of inherent interest to geographers, because they vary so greatly from one tobacco producing district to another, and form such prominent landscape features in each district. These areal variations have an intimate functional relationship with important differences in the kinds of tobacco produced in the various producing districts. Tobacco barns may be of considerable value in providing a ready clue to the character of the entire rural economy of a tobacco producing area.

The name "tobacco barn" is applied to a surprising variety of structures in different parts of the United States. It is used to refer to structures devoted entirely to tobacco curing; to structures used for curing tobacco but which also serve as general farm storage units; to structures with an upper portion louvered and arranged for tobacco curing but with a ground floor for stabling horses; to structures which are used to cure tobacco at one season of the year, and to provide sheltered feed yards for cattle at another; to structures which stand separate from "stripping" sheds as well as to those which have provision for "stripping" within, beneath, or in rooms attached to the building; to structures incorporated in the general cluster of farmstead buildings as well as to those set in the midst of fields or half-hidden in nearby woods; to log barns, to dilapidated sheds, and to huge structures set against a background of vast fields shaded with cloth. Further differentiation is provided by regional variations

* "The Character of Tobacco Barns and their Role in the Tobacco Economy of the United States," with Eugene Cotton Mather, *Annals of the Association of American Geographers* 51 (1961): 274–93.

Table 6.1 United States Tobacco: Classes, Types, and Principal Uses

Class	Type	Principal Use
1. Flue-cured	11. Old Belt	Cigarettes, export, pipe tobacco, plug wrapper and filler
	12. New Belt	
	13. Carolina Border Belt	
	14. Georgia-Florida Belt	
2. Fire-cured	21. Virginia	Export, snuff, plug wrapper and filler, Italian-type cigars
	22. Eastern	
	23. Western	
	24. Northern	
3. Air-cured	31. Burley	Cigarettes, pipe, plus
	32. Maryland	Cigarettes, export
Dark Air-cured	35. One-sucker	Plug, snuff, export, pipe tobacco
	36. Green River	
	37. Virginia Sun-cured	
4. Cigar Filler	41. Pennsylvania Seedleaf	Cigar filler
	42. Gebhardt	
	43. Zimmer	
	44. Dutch	
	46. Puerto Rico	
5. Cigar Binder	51. Connecticut Broadleaf	Cigar binder
	52. Connecticut Havana Seed	
	53. New York and Pennsylvania Havana Seed	
	54. Southern Wisconsin	
	55. Northern Wisconsin	
6. Cigar Wrapper	61. Connecticut Shade	Cigar wrapper
	62. Georgia-Florida Shade	
7. Miscellaneous	72. Perique	Pipe tobacco

After Charles E. Gage, *American Tobacco Types, Uses, and Markets*. U.S. Department of Agriculture, Circular No. 249, 1942.

in the favorite color—or the lack of it—of tobacco barns: red in Wisconsin, white in Pennsylvania, unpainted in Maryland and Tennessee, and two–tone in Kentucky, where vertical ventilator boards are painted in contrast to the rest of the barn.

Despite variations in external form, however, the interiors of most tobacco barns are similar, with a lattice of tier poles on which tobacco can be hung to cure. Tobacco is attached to sticks before it is hung in the barn; because all tobacco sticks are about four and a half feet long, the barn has a basic framework of horizontal tier poles spaced four feet apart horizontally and twenty inches to five feet vertically. The lowest tier pole is six to nine feet above the barn floor. Tier poles are supported by transverse lines of posts and crossbeams, and the space between transverse lines is known as a "bent." Tobacco farmers commonly think of barn size in terms of bents, and in some areas they use the rule of thumb that one bent will hang half an acre of tobacco.

Green tobacco, when it is brought into the barn, has a water content of eighty to ninety percent, and the basic purpose of the curing process is to remove this water in such a way as to facilitate those chemical changes which will produce leaf of the desired character. Although the basic characteristics of the leaf have been fairly well established by the time it is harvested, the farmer can ruin a good crop by curing it improperly, and he cannot market the crop until it has been cured. In order to control the curing process, of course, he must have the right kind of barn. Areal variations in tobacco barns thus reflect, to a considerable degree, the localization and specialization of commercial tobacco production in this country.

Although only one single botanical species of tobacco (*Nicotiana tabacum*) is grown commercially in the United States, the tobacco trade—from grower to manufacturer—recognizes some twenty-six different types of tobacco, as designated by the U.S. Department of Agriculture (Table 6.1). Each type is produced in a fairly well-defined area and is especially suited to the manufacture of one or more particular tobacco products, but it cannot be used at all in the manufacture of others. The desirable characteristics of the leaf result primarily from the climatic and soil conditions of the producing district, but these characteristics can be enhanced by use of seed of certain varieties, special methods of fertilization, cultivation, harvesting, and proper curing.

The twenty-six different types of tobacco are grouped into seven major classes on the basis of method of curing, method of marketing, and ultimate use

(Table 6.1, Fig. 6.1). The "southern" tobacco of Classes 1, 2, and 3 is grown in the former slave states of the South, sold at auction sales in warehouses located in urban areas, and used in the manufacture of all forms of tobacco products except cigars. It is classified on the basis of method of curing. Tobacco of Class 1 (Flue-cured Types) is hung over metal flues and cured by radiant heat which reaches temperatures as high as 180°F. Tobacco of Class 2 (Fire-cured Types) is hung over a smouldering fire of hardwood and sawdust, and absorbs a distinctive aroma from the smoke. Tobacco of Class 3 (Air-Cured Types) is cured in barns with controlled ventilation, and heat plays no role in the curing process, although it may be used to control humidity in damp weather.

The "northern" tobacco of Classes 4, 5, and 6 (Cigar Types) is grown mainly in the old free states of the North, is air-cured, sold directly to buyers who visit individual farms, and used almost exclusively in the manufacture of cigars. It is classified on the basis of its major use. Tobacco of Class 4 (Cigar Filler) is used in the core, which comprises the major part of the cigar. Single leaves of tobacco of Class 5 (Cigar Binder) are bound around this core of filler tobacco to hold it together, and single leaves of tobacco of Class 6 (Cigar Wrapper) are wrapped around the binder to form the outer wrapping which gives the cigar its characteristic appearance.

Tobacco Barns in Western Kentucky and Tennessee

Western Kentucky and Tennessee have two or three different kinds of tobacco barns because this is an area where traditional patterns of tobacco production have been modified greatly by the "Cigarette Revolution." Most of the oldest barns in the area are used for fire curing. These are comparatively small, unpainted structures of frame or log, perhaps twenty to thirty-two feet wide and twenty-six to forty-eight feet long, but high enough for five sets of tier poles spaced three and a half feet apart. Two large doors at each end permit loaded tobacco wagons to be driven through the barn. Once the barn has been filled with tobacco, the fires of hardwood, smothered in damp sawdust, are laid right on the wagon tracks. The barn should be airtight, although few of them actually are, and ideally it should be located in a low place near woods for shelter and maximum heat efficiency. Although some fire-cured tobacco barns in this area are no longer used, the casual observer might overestimate the degree of abandonment, because so many barns are set back in the woods and appear relatively unkempt.

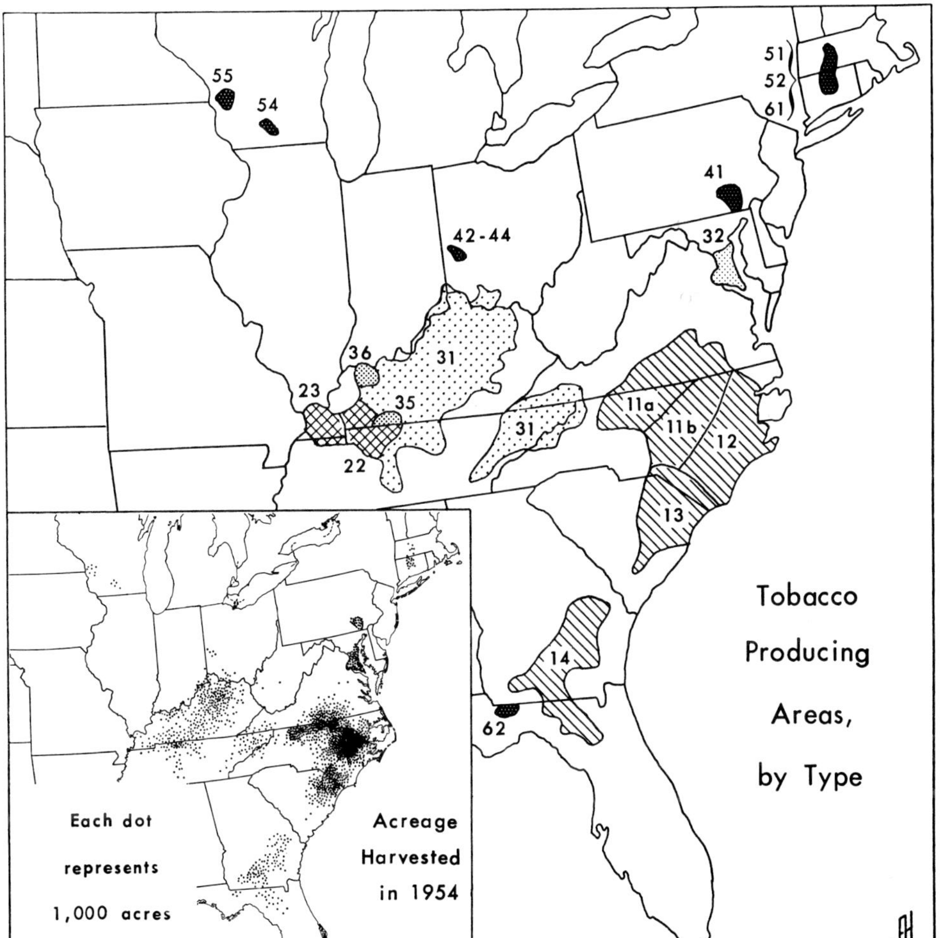

FIG. 6.1. Tobacco producing areas by type in 1954; see Table 6.1 (p. 74) for an explanation.

The barn used for air-curing dark tobacco is unlike the fire-cured tobacco barn and rather similar to other types of air-cured tobacco barns, albeit somewhat smaller. Low, dark tobacco plants require only 1.25 to 1.5 cubic feet of barn space apiece, whereas the taller Burley plant requires two to 2.5 cubic feet, and some farmers argue that a barn which holds an acre of Burley will hang three acres of dark tobacco. The barn built for air-curing dark tobacco is most commonly an unpainted frame structure with a two-foot-wide vertical ventilator door for each twelve-foot bent. It may have a line of ventilators along the peak of the roof, and commonly has only one door at each end of the barn. It also differs from the fire-cured tobacco barn in location, for the air-cured barn is placed in the most exposed position available in order to obtain full benefit of wind from any direction.

The dark tobacco—both air and fire cured—of western Kentucky and Tennessee is closer than any other modern tobacco to the original type which was planted at Jamestown in 1612. Early settlers were producing tobacco in this area by 1800, and for more than a century it was a staple export commodity. Smoke was used in the curing process, both here and in colonial Virginia, to improve the keeping qualities of the tobacco while in transit to Europe, and this area continued to depend on the export market long after Eastern tobacco districts had turned their sights onto the domestic consumer. The world demand for dark tobacco dropped sharply in the 1920s, however, and domestic demand for snuff and chewing tobacco—the only tobacco products in which large quantities of dark tobacco are used—has also steadily declined. As a result, production of dark tobacco has plummeted in the last forty years. In recent years, dark-fired (Types 21-24) and dark air-cured (Types 35-37) tobacco have had some of the lowest per acre returns of any tobacco types.

The heart of the dark-fired tobacco country is in the western part of the two states, around Mayfield and Murray in Kentucky and Clarksville and Springfield in Tennessee. Dark air-cured tobacco, which is grown between this area and the main air-cured Burley area of central Kentucky, appears to be encroaching on the dark-fired tobacco area as farmers switch to the easier curing system, and the more profitable Burley has definitely been invading areas where formerly dark air-cured tobacco was produced. A farmer who switches from dark-fired to dark air-cured tobacco often has to build a new barn, and the old one stands relict, but Burley can be cured just as easily as dark tobacco in a barn built for air-curing.

FIG. 6.2. A large air-cured tobacco barn, with the farmer standing in the open ventilator to give and indication of scale.

Burley Tobacco Barns

Burley tobacco barns are archetypical of air-cured tobacco barns in the United States (Fig. 6.2). They commonly are twenty-eight to forty feet wide and may be up to 300 feet long, although the majority are between thirty-six and seventy-two feet long. The height to the eaves may be anywhere from sixteen to twenty-four feet. All sideboards are vertical, and every third board is hinged on one side so that it may be opened for ventilation. A driveway extends lengthwise through the barn, with doors at either end large enough to admit a truck or wagon loaded with tobacco. Although heat plays no role in the curing process, small fires may be built on the floor of the barn in damp weather to help control humidity.

If possible, the barn is located on a ridge or hilltop, rather than in a valley, in order to take advantage of wind from any direction. The stripping room or shed, where cured leaf is "stripped" from the stalk, is attached to the barn about midway along one side to avoid unnecessary carrying of tobacco. The stripping shed normally extends in an east-west direction so that the stripper is facing north and examining the tobacco by indirect daylight; variable cloudiness can influence his evaluation of leaf color so greatly that he dares not strip while facing the sun.

A considerable degree of variety characterizes the individual barns of the Burley tobacco-producing area, which includes large portions of Kentucky and Tennessee and smaller areas in western North Carolina, southwestern Virginia, and southern Indiana and Ohio. The three principal areas of production are the Bluegrass country in north-central Kentucky, the Pennyroyal karst country south and east of Mammoth Cave along the Kentucky-Tennessee line, and the valleys and mountains of northeastern Tennessee. In each of these areas Burley tobacco grows best on rich silt-loam soils derived from limestone, and many tobacco farmers argue that Burley tobacco is strictly a limestone country type.

In the Kentucky Bluegrass area, most tobacco barns are painted, and many have ventilators painted in contrast to the rest of the barn. Black barns with white ventilators, for instance, are common, as are white barns with green ventilators. Painted barns are indicative of the general agricultural prosperity of this area, where intensive tobacco farming is associated with extensive livestock operations. Many of the livestock farms are showplaces where fine race horses are bred, but even the fanciest horse farms have a field of tobacco and a tobacco

FIG. 6.3. Stripping tobacco in a Kentucky Pennyroyal barn. The workers break the cured leaves from the stalk, grade them, and tie them into bundles of a single grade.

barn tucked away at the back end of the farm. The two operations are quite openly combined on less pretentious horse farms, and the barns are used both for curing tobacco and for stabling horses.

In the Pennyroyal karst country, tobacco barns are not quite so imposing as in the Bluegrass country. Farms are smaller and less prosperous than in the Bluegrass, so there is less tobacco to be put into the barns, and less income with which to buy paint for them. Pennyroyal tobacco barns, therefore, are smaller than those of the Bluegrass, and the great majority are unpainted frame structures. Many farmers in this area manage to get along without separate stripping sheds by working on temporary tables of sawhorses and boards set up in the barn (Fig. 6.3).

Barns of Other Air-Cured Tobacco Areas

All northern, or cigar, tobacco is air-cured, as is the tobacco produced in the southern counties of Maryland. The production and curing of cigar wrapper tobacco (Class 6) is so unusual that it requires separate consideration, but the tobacco barns of southern Maryland, southeastern Pennsylvania, southern Wisconsin, and the cigar binder tobacco barns of the Connecticut Valley are all fairly similar to the barns in which Burley is cured.

A great variety of roof profiles is a striking feature of tobacco barns in southern Maryland. The majority of roofs are of the common gable type, with symmetrical slopes (Fig. 6.4), but one also finds asymmetrical gable roofs with one side steeper than the other, hip roofs, and concave upward shed roofs formed by extending a level roof outward from the bottom of a common gable roof. Although these roof variations appear designed to increase barn space, the value of additional space may be offset by the peculiar ventilation problems created.

The tobacco barn of southern Maryland is located near the tobacco field and the farmstead, and preferably on a hilltop for good ventilation and drainage of air and water. Very few are painted. Some of the older barns are ventilated only by access doors and loose construction, but the more modern examples have vertical ventilators hinged either at the sides or, more commonly, at the top. Tobacco may be stripped in a separate building adjacent to the barn, but many farmers have built stripping rooms inside the barn itself. They find that they can retain more heat and moisture by excavating to a depth of three or four feet and constructing a semi-basement stripping room.

FIG. 6.4. Air-cured tobacco barn in southern Maryland.

FIG. 6.5. A five-acre, air-cured tobacco barn in southern Wisconsin.

Tobacco barns are fairly standardized in the two areas of southern Wisconsin where cigar binder (Types 54 and 55) tobacco is produced. They are twenty-eight feet wide, fifteen feet to the eaves and twenty-five to the gable, and as long as necessary to hang all the tobacco grown on the farm. The barn is of frame and board construction, and most commonly is painted red with white doors (Fig. 6.5). In the southern (Type 54) district south of Madison, centering on Stoughton and Edgerton, the tobacco barns commonly are in or adjacent to the farmstead, and they are located with respect to the effect of topography and trees on air circulation. The northern (Type 55) district is centered near Viroqua in the Driftless Area, and here many barns are situated in valleys because there is no alternative location for the cultivated land and the farmstead.

Production of shade-grown cigar wrapper (Class 6) tobacco is both spectacular and expensive. Even its history is of more than usual interest. In the last century, most domestic cigars were wrapped with tobacco from the Connecticut Valley, but after 1880 the native leaf began to face serious competition from a far superior wrapper leaf imported from Sumatra. Connecticut growers obtained a protective tariff against Sumatra tobacco, but they were distressed in 1896 to learn that tobacco grown in Florida under shade—to simulate the climate of Sumatra—was almost indistinguishable from Sumatra leaf. Shortly thereafter, however, soil surveys showed that sandy loam terrace soils of the Connecticut Valley were remarkably similar to those on which Sumatra type leaf had been grown successfully in Florida, and within half a dozen years the production of shade-grown cigar wrapper tobacco had become firmly established in Connecticut.

All cigar wrapper tobacco produced in the United States is grown under cloth shades which protect the plant from direct sunlight and tend to simulate a moist tropical climate by increasing humidity and reducing evaporation and wind velocity. Specially prepared cotton cloth is sewn to galvanized wire stretched between red cedar posts set in precise thirty-three-foot squares. The twelve-foot posts are set three feet in the ground, so that the cloth is eight to nine feet above the surface of the fields. The cloth is used on top only one year, but some of it is used a second year in double thickness on the sides. At both ends of the field, the sides of the shade are arranged so that they can easily be raised for such essential operations as planting, cultivation, tying, and harvesting.

Unlike other air-cured tobacco types, shade-grown tobacco is harvested by "priming." The farmer has each individual leaf picked as it becomes ripe, rather than cutting an entire plant which has leaves in varying stages of maturity. The field is picked every week or ten days, and two to four leaves are taken each time. The primed leaves are placed in baskets and carried to the barns, where they are sewed onto sticks and hung for curing. The usual shade-grown tobacco barn is forty feet wide, 100 to 140 feet long, and eighteen to twenty-two feet to the eaves. The tier poles are spaced vertically at intervals of only two to two and a half feet, because only the leaf is cured. Ventilators on the side of the barn are located so as to be open between the tiers of tobacco inside and thus permit unimpeded air movement (Fig. 6.6).

Production of shade-grown tobacco in the Connecticut Valley is largely in the hands of corporations, for very few individual farmers have the necessary capital. A corporation often has fifty to sixty acres of tobacco in a single field, and fifteen to twenty barns may be clustered on one side to cure the tobacco from this one field. Nearby is a dormitory, for much of the work in the fields and barns is done by school boys during their summer vacations. The workers come around the Fourth of July, in time to tie the plants to the overhead wires, and stay until mid-August, when most of the picking, sewing, and hanging have been completed.

Flue-Cured Tobacco Barns

Flue-cured tobacco barns are small, square, and distinctive (Fig. 6.7). They measure sixteen to twenty (rarely twenty-four) feet on a side, and sixteen to twenty feet to the eaves, with two small access doors on opposite sides of the barn. Inside, the tier poles start about six feet above the ground, and are spaced at vertical intervals of two to 2.5 feet, so that a barn normally contains six or seven vertical tiers. A three-foot square brick furnace squats at one side of the barn, and from it extend two flues of twelve-inch sheet metal pipe which cross the barn, make two right angle bends, and emerge from the side of the barn on which the furnace is located. These flues have smoke-stacks to discharge gases and increase the draft.

The side of the barn with the furnace commonly has a lean-to shelter or shed to protect the furnace, the fuel supply, and the attendant. Wood formerly was the common fuel, and required a larger furnace, but today fuel oil, gas, and coal are all used. Stoker feeding and thermostatic control provide more uniform

FIG. 6.6. Horizontal ventilators on air-cured, shade-grown tobacco barns in the Connecticut River valley.

FIG. 6.7. A flue-cured tobacco barn in the New Belt of eastern North Carolina. Workers under the shed are stringing primed leaves which have just been brought in from the field.

curing conditions and require less supervision. The fuel supply—and stoker, if one is used—normally are visible under the shed, which also provides shelter for stringing operations at the time of harvest.

All flue-cured tobacco is harvested by priming, or picking each leaf as it becomes ripe. The harvest is labor hungry. The field is picked about once a week for six to eight weeks, and each leaf is handled individually at least four times. Pickers walk down the rows and pull off all ripe leaves, which they place in mule-drawn sleds with waist-high burlap sides; the rows are so close together that tractors would damage the plants. Mules haul the full sleds to the shelter of the barn shed, where a "hander" picks up several leaves at a time and arranges them into a neat bunch before passing them to the "looper," or "stringer," who ties them to tobacco sticks with loops of string. The sticks are then hung in the barn for curing. Each picking is cured and taken from the barn before the next picking is made. After the leaf has been cured it must be untied from the tobacco stick and sorted into grades before it is ready for market.

A barn of flue-cured tobacco normally is cured in 3.5 to five days. The barn is filled in a single day, and the furnace is started. In the first stage, which lasts twenty-four to thirty-six hours, the temperature is kept between 90º and 100ºF to change the green leaf to the desired lemon-yellow color. Then the temperature is raised to 130º–140ºF for another twenty-four to thirty-six hours to dry the leaf before its color changes to reddish brown, and finally the temperature is gradually increased to 170ºF or 180ºF to dry the stem. The cured leaf has a bright yellow-orange color, hence the name "Bright" which is commonly applied to flue-cured tobacco.

Historically, flue-cured tobacco is the product of change in the gross characteristics of the tobacco plant which occurred as a result of the change in soil type from the heavy loams of central Virginia to the less fertile, light gray soils of the North Carolina border counties. Seeds of the heavy leaved, dark green plant grown in Virginia produced a plant with thin, light green leaves when grown under conditions of semi-starvation on infertile soils of the North Carolina Piedmont. An even more desirable leaf was produced when semi-starvation was continued in the flue-curing process, which uses high temperatures to secure rapid dehydration and inhabit metabolism in the leaf.

Flue-cured tobacco was initially used as a wrapper for plugs of chewing tobacco because of its attractive bright color, but its real popularity has been associated with the Cigarette Revolution. Although the precise blend of tobacco used in any cigarette is a guarded trade secret, the average cigarette probably contains fifty-three percent flue-cured tobacco, thirty-three percent Burley, four percent Maryland, and twenty percent oriental (from Turkey and Greece). The popularity of flue-cured tobacco is indicated both by its increase from a third of the nation's tobacco acreage and a quarter of its production in 1910–1919, to more than three-fifths of both in 1953–1957.

Flue-cured tobacco was first produced along the Virginia-Carolina border in the Old Belt where Type 11 tobacco is grown today. The tobacco barns of the Old Belt are distinctly older and smaller than barns in other flue-cured tobacco districts. From the Old Belt, production of flue-cured tobacco has expanded into new areas, mainly at the expense of cotton. In the late 1880s, for instance, when cotton prices dropped disastrously, many farmers in eastern North Carolina began to grow tobacco. They retained the traditional system of tenancy with the new crop, however, and much of the tobacco of eastern North Carolina is still produced by African-American share croppers. Unlike the yeoman farmers of the Old Belt, the tobacco farmers of the New Belt are prodigal of manpower. A minimal harvest crew consists of four primers, four handers, two loopers, and a sled driver. Farmers in this area have expressed considerable interest in a machine harvester which, despite its almost unbelievable cumbersomeness, reduces the minimum harvest crew size from eleven workers to seven. It is noteworthy that rural farm population densities in eastern North Carolina tobacco areas were the densest in the nation in 1950.

The change in soils from the Piedmont to the Coastal Plain was responsible for a new type of tobacco (Type 12), which is grown in the area known as the New Belt. Most of the tobacco barns in the New Belt are sheathed in insulating material covered by green or black tarpaper, which is held in place by thin vertical wooden battens nailed to the lumber of the barn. An adjustable ventilator may run the length of the ridge line, and many barns have a smokestack jutting from each corner of the roof. Shelters are constructed on only one side, although often a shelter connects two barns. Fuel oil and gas have virtually replaced wood as fuel since 1940, but many barns are still built near the edge of the woods to reduce heat loss on windy days.

Further Observations

Our primary concern here has been the description of regional variations in tobacco barns, with some analysis of the relationship between the character of the barn, its site, and its role in the tobacco economy of the region and the individual farm unit. Both flue-cured and fire-cured tobacco barns are distinctive, and their distinctiveness, to a large degree, is functional; they are specifically designed to cure one single class of tobacco, and cannot be used to cure tobacco of any other class. The relationship between form and function is also noteworthy in the case of air-cured tobacco barns, although it is complicated by regional variations in style which are more closely related to the culture of the region than to the function of the barn. Like fences, tobacco barns and other human-built elements of the agricultural scene often require us to extend our attention beyond the farmstead itself.

Land-Use Change

CHAPTER 7

in a Piedmont County*

The Piedmont area of Georgia and the Carolinas lost farm land at a truly spectacular rate between 1939 and 1974. Nearly ten million acres of land, an area one-fourth the size of the entire state of Iowa, passed out of farmer ownership in eighty-nine counties between Charlotte, North Carolina, and Birmingham, Alabama, within a third of a century. Roughly two-fifths of this land was wooded, and presumably had contributed only minimally to farm production, but the loss of four and a half million acres of harvested cropland plus one and a third million acres of other cleared farm land is genuinely staggering.

The decline in cotton acreage dramatizes the change in these eighty-nine Piedmont counties. They had sixty percent more land in cotton alone in 1939 than their acreage of all cropland harvested in 1974, even though their cotton acreage had already dropped to half its 1929 peak of 2,871,000 acres by 1939. The southern Piedmont had been the first major upland cotton-producing district in the South, and it remained a major cotton-producing district until World War II, but since 1950 cotton has virtually vanished from the region, and no other crop has come along to take its place.

The acreage of cropland harvested has dropped most massively in the southwestern Piedmont, especially in the area around and west of Atlanta, which has somewhat rougher topography than the rest of the region. Some folk might be quick to invoke urban sprawl to explain these losses near Atlanta, and there can be no doubt that metropolitan expansion has indeed removed some

* "Land Use Change in a Piedmont County," *Annals of the Association of American Geographers* 70, (1980): 492-527.

land from agricultural production, but such a glib explanation is challenged by two simple facts: (1) the urban/industrial areas of the Carolina Crescent [the concentration of cities between Raleigh, North Carolina, and Greenville, South Carolina] are growing just as rapidly as the southwestern Piedmont, but have lost far less cropland; and (2) losses on the southwestern Piedmont have been just as heavy, or even heavier, in areas remote from Atlanta, which is the only major urban center in the region.

The champion loser of cropland between 1939 and 1974 was Carroll County, Georgia, which dropped from 129,927 acres of cropland harvested in 1939 to only 7,801 acres in 1974. This loss was equivalent to two-fifths of the county's total area, and by 1974 crops were harvested from only one acre of every forty in the county. Carrollton, the county seat, is some forty miles west of Atlanta, and the county presumably is at or even beyond the outer margin of daily commuting to the metropolis, although it is nestled between two spokes of the Interstate highway system that radiates from Atlanta.

What Happened to Cotton?

Local people like to boast that Carroll County was the second largest cotton-producing county in Georgia before it was hit by a double plague of boll weevils followed by bureaucrats. Production dropped from 34,000 bales in 1920 to 26,000 bales in 1921, when the boll weevil arrived, but bounced back to better than 30,000 bales a year in the mid-1920s and exceeded 40,000 bales a year in 1927. The bureaucrats did more damage in 1933 when they reduced the county's cotton allotment from 63,000 acres to 44,000 acres under the New Deal price support program.

The history of land use change in Carroll County can be envisioned by tracing an imaginary piece of land. In the 1920s, the farmer grew cotton on this land, but the bottom dropped out of the price of cotton during the Depression. After 1933, the government restricted the acreage he could grow, so he began using his old cotton land to grow fodder crops, such as corn, hay, oats, winter wheat, or soybeans (which were cut green for leguminous hay, not harvested for beans). Then along came World War II; the boys all marched away, and the farmer had so much trouble finding the help he needed to grow crops that he turned his cropland into pasture. He himself got a job in town, and gradually he began to lose interest in his pasture land. He allowed it to grow up in brush and second-growth hardwood, or he sold it to a paper company to be planted in loblolly pine.

The distribution of the remaining agricultural land and the built-up land in Carroll County in 1974 reflected the physical character of the land surface (Figs.7.1 and 7.2). Most of the county's nonwooded land was concentrated in a broad northeast-southwest swath on the gently rolling central upland that is a relatively undissected remnant of the Piedmont peneplain. The southeastern and northwestern corners of the county have been deeply fretted by streams, and the agricultural land in these rougher and more dissected areas was largely restricted to strips along the narrow, crooked, and uneven ridge crests, which are all that remains of the former peneplain.

Most land in Carroll County, except for the poorly drained bottomlands, probably has produced a crop of cotton at one time or another. The cotton plant "does not like to get its feet wet," so the bottomlands have been used for corn or pasture, or left in hardwood forest, but cotton probably has been grown on every acre of upland in the county, including some surprisingly steep slopes. Today, when two-thirds of the county is wooded, it is hard to believe that some farmers had to buy firewood in the late 1930s because they had so few trees on their land.

Cotton was hard on the land, especially when it was not followed by a winter cover crop. Much of the topsoil of the uplands has paid the price of long-continued row crop agriculture, and contemporary farmers are tilling the stony subsoil. By the 1920s, most farmers had ridged their fields with low earthen contour terraces that did not interfere with cultivation, but slowed down runoff and thereby reduced the rate of erosion. Far more terraces were constructed than were actually needed, because the U.S. Soil Conservation Service subsidized them heavily, they were easy to construct in the off season with farm equipment and farm labor, and they posed no problems back in the mule days. Today, these terraces wind through pastures and wooded areas as well as across plowed fields, and are a splendid indicator of the former extent of cultivated land.

Most old-timers in Carroll County, when I asked them why they had stopped growing cotton, gave me some variant of the response that they "couldn't make it pay." They were caught in a bind of declining soil fertility, decreasing yields, falling cotton prices, and increasing costs, especially wages. Traditional cotton farming demanded a large supply of cheap labor for the twin chores of "chopping" weeds out of the young crop and picking the ripened bolls. The

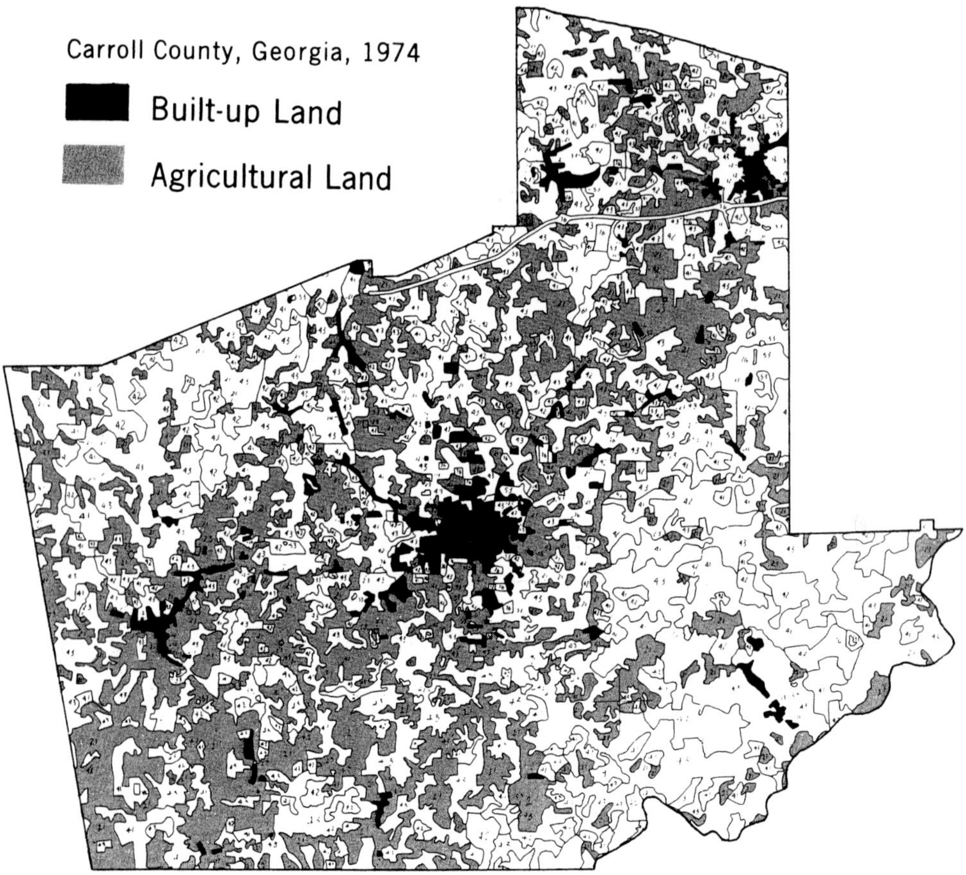

FIG. 7.1. Agricultural land and built-up land in Carroll County, Georgia, 1974.

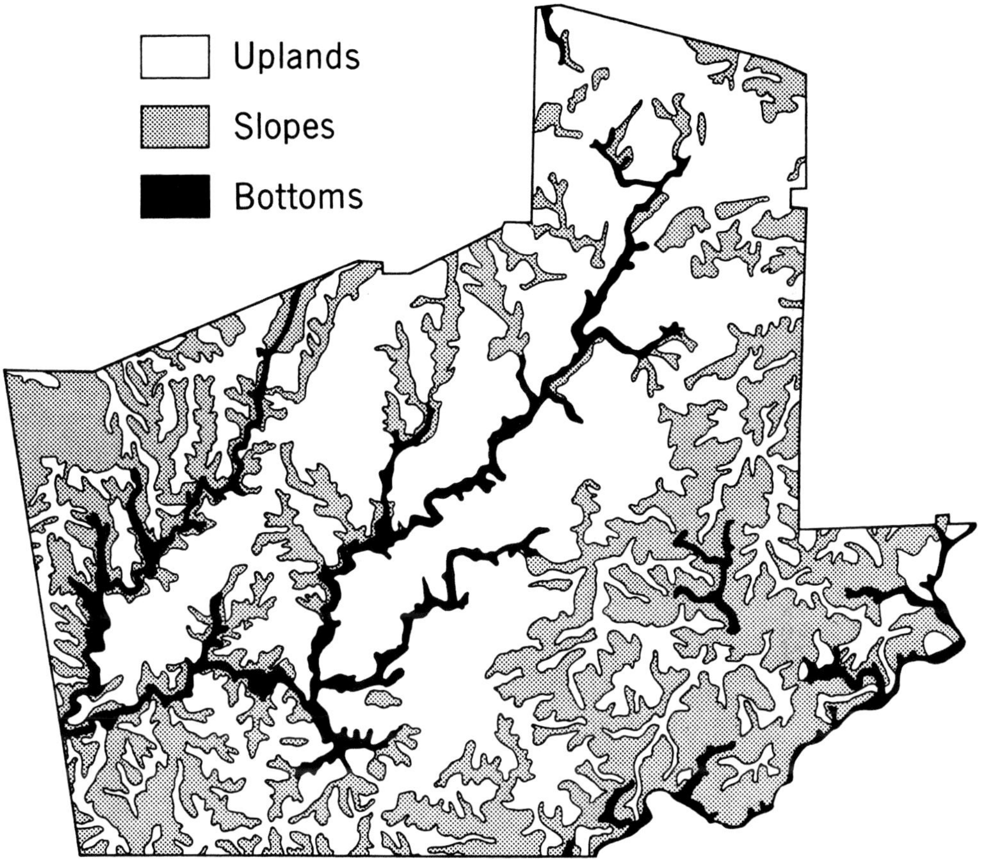

FIG. 7.2. Land surface features of Carroll County.

harvest season was especially demanding, and most farmers had to hire extra help for the back-breaking stoop labor of cotton picking. Some stopped growing cotton when they could no longer afford to pay for having it picked, others when they realized that their own labor was worth more in town than it was on the land.

In some areas, the answer to rising wages was a changeover from men and mules to machinery, but Carroll County had never enjoyed the large plantation landholdings that facilitated such a changeover. The county had too many undersized "forty acres and a mule" cotton farms that were too small to afford machines and too broken topographically to use them efficiently. In the 1920s, the modal size of landholding in Carroll County was around 100 acres, which was roughly twice as much as one man and his family could handle with a team of mules. Part of nearly every landholding was farmed by sharecroppers or tenants. Many cropper units were one-mule operations of only twenty acres or so that were too small to afford a tractor when tractors became available.

The county tax records for 1978 reflect the heritage of small yeoman cotton farms. Only thirty-four owners paid taxes on 320 acres or more. At the other extreme, an astonishing fifteen percent of the county was in rural properties of less than forty acres, and more than thirty percent was in rural properties of less than sixty acres, far too small to provide a reasonable level of income from any but the most intensive and highly specialized agricultural activities.

Three Who Have Succeeded

The person who wished to continue farming in Carroll County faced the extraordinarily difficult task of assembling enough small property units to comprise a farm of reasonable size. It is always easier to buy one large tract of land than many small tracts that add up to the same size. After someone has bought a few tracts, the owners of the rest figure out what he is up to, and they jack up the price; some cantankerous owner right in the middle may flatly refuse to sell at any price. Only a handful of Carroll County farmers have had the imagination, entrepreneurial flair, managerial skill, and capital necessary to enlarge their operations and continue farming on a full-time basis.

Larry Staples, Ned Harman, and M. F. Peek are examples of the few who have succeeded. Mr. Staples has developed old cotton land into a dairy farm of 800 acres, and he milks 325 cows. He has 450 acres of cropland on which he raises wheat in winter and no-till [a culture that eliminates turning over the soil with a plow] corn in summer. Both crops are cut for silage. Mr. Harman has a farm of 990 acres, with 460 acres of cropland on which he raises cash crops of corn and soybeans. He plants wheat and rye on his cropland for winter cover, and uses it to graze a herd of 400 beef cattle. He also has 250 brood cows, and in 1979, after two consecutive years of severe drought, he bought a center pivot sprinkler irrigation system. Staples and Harman are two of the larger landowners in Carroll County, and between them they planted almost twelve percent of the total acreage of cropland harvested in the entire county.

Mr. Peek has an old cotton farm of 170 acres, of which 100 acres are in pasture; he has no cropland. He has a breeding herd of forty beef cows, and fattens broilers and hogs in long, low houses that are specially designed for that purpose. Mr. Peek is also a large landowner by Carroll County standards, but he is closer to the county norm because he has converted all the old cotton land on his farm into pasture, and because he produces broilers, which provide three-fourths of all farm income in the county. He differs from most other poultry farmers because he makes a living at it, and most poultry farmers are part-time operators. He is able to make a living at it because he has enough land for a beef cattle operation that is large by county standards, but he has also had to add a hog operation, and he told me that he has had to refinance his land from time to time in order to meet his operating expenses.

What Are the Options?

An American farmer who can no longer engage in a traditional farm activity has surprisingly few options for maintaining his former level of economic return even if he has an unlimited amount of land. The options are even further constrained by the fact that the United States currently has far more farm land than it needs to produce all the food and fiber the American people can consume—in recent years, the nation has exported the production of one of every three harvested acres. Large areas of mediocre to poor farm land simply are not needed, at least for the time being, to satisfy the nation's requirements, and they have passed out of agricultural production.

This situation obviously could change, and today's abandoned farm land could easily become tomorrow's priceless asset. The abandoned cotton land of the Piedmont, which has little place in the contemporary agricultural economy, might well be an acreage reserve of potential cropland that is recouping its fertility under hardwood forest, but such a though is scant solace for farmers who must make a living from the land today.

Broilers

Poultry was far and away the most important source of farm income in Carroll County in 1974. Estimates of the number of poultry farmers in the county range from sixty-five to 140. Almost anyone who can borrow the money to build a broiler house can get into the poultry business, and getting out is just as easy. Long, low, one-story broiler houses are scattered all over the country, commonly in nests of two or three. Each house has its own automatic feeding and watering system, with a pair of shiny metal tanks for feed storage.

The production of broilers has been an attractive option for many small farmers, especially those who have off-farm jobs, because it produces regular income, requires only three to five hours a day, and much of the work can be done by women and children. Broiler production has been singularly popular in areas where farmers could no longer rely on a previously important source of income, such as cotton in Carroll County. These areas have developed the infrastructure necessary for the broiler business. Local banks will finance construction of broiler houses, and local builders know how to put them up. Feed companies provide chicks, feed, close supervision, and a guaranteed price.

The broiler business is controlled by large feed or poultry companies, such as Gold Kist, Inc., which have made truly remarkable improvements in feed conversion and fattening time. Between 1957 and 1979 the amount of feed necessary to fatten a bird was cut by twenty percent, and feeding time was halved, but the price remained almost the same. The improved technology has put pressure on small poultry farmers because the poultry companies do not like to deal with farmers who can house fewer than 12,000 birds. Many poultry farmers, in turn, intensely dislike the close supervision involved in broiler production, because no farmer likes being told what to do on his own farm. Some broiler farmers mutter that they are little better than sharecroppers, but that regular broiler check still looks pretty good to them.

Woodland

The most widely exercised option on former cotton land in Carroll County, as in so much of the South, has been simply to leave it alone and let it grow up in weeds, brush, and trees (Fig. 7.3). In 1935, woodland covered slightly more than a third of the county, and most of it was owned by farmers; by 1974, the woodland acreage had nearly doubled, but most of it had passed out of farm ownership and was in small, private woodlots that were managed poorly, if at all. Some landowners, perhaps many, still think of trees merely as weeds that colonize old farm land, and they have little interest in good forest management. The original upland forest in Carroll County was a mixture of oak and pine, and this kind of forest quickly reestablishes itself on abandoned farm land. The principal objective of contemporary forest management on the Piedmont is to produce fiber forests of loblolly pine by controlling or eliminating the less valuable hardwoods.

Approximately half of Carroll County is covered with hardwood or mixed hardwood and pine forests that are not being managed efficiently for production of forest products. A well-managed forest would be harvested once every twenty to thirty years. A large landowner could ensure a reasonably steady income by having part of his land in each stage of the production cycle, but the small woodlot owner has too little land to subdivide into twenty to thirty parcels, too little money to hire a trained forester to manage his land, and too little patience to wait half a lifetime for a return on an investment in planting and maintaining pine trees.

Foresters have already started to fret about the spectre of a potentially severe wood shortage after the present "bonanza" pine forest has been harvested. Much of this forest is growing on former farm land that was abandoned in the 1950s. Some Carroll County landowners were quite happy to sell their land to pulp and paper companies in the early 1950s, although these companies were less active in the county than they were in other parts of the South.

The pine forest that has grown up on old cotton lands has been a bonanza for pulpwood manufacture, which has been one of the most vigorous growth industries of the South since World War II. In 1950, the region had fifty-five pulpwood mills with a total capacity of 23,197 tons per day; in 1976, it had 115 mills with a capacity of 101,513 tons per day. The amount of land required to keep these mills running smoothly can be estimated from the fact that a cord

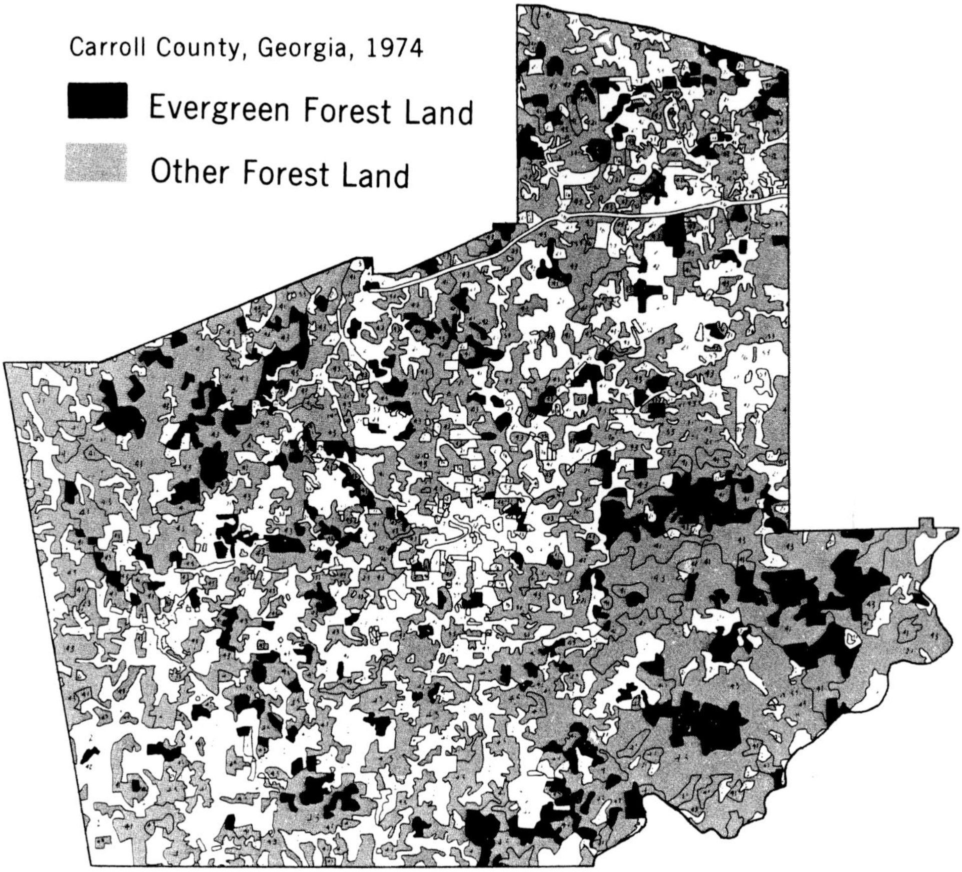

FIG. 7.3. Forest land in Carroll County, 1974.

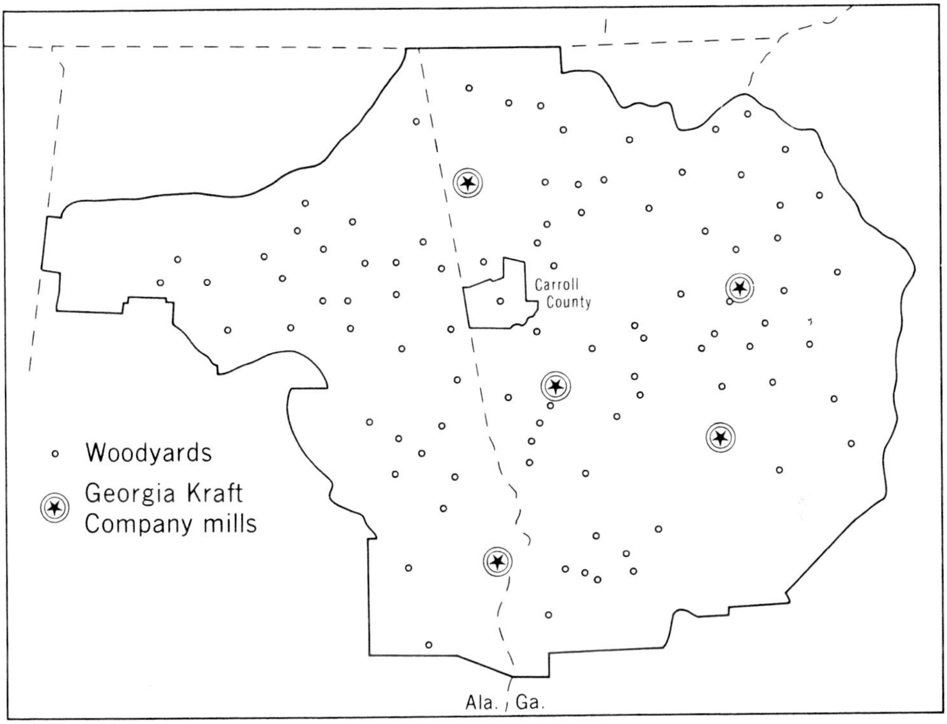

FIG. 7.4. Wood yards and paper mills in western Georgia and eastern Alabama, 1978.

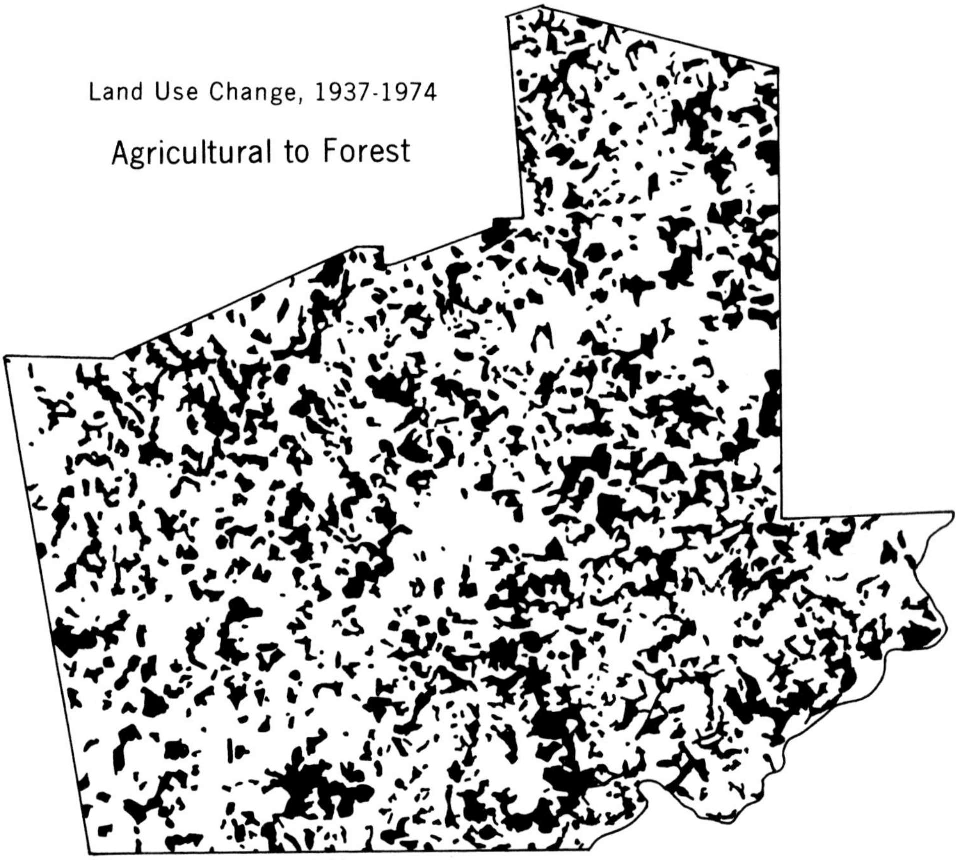

FIG. 7.5. Conversion of Agricultural to forest land in Carroll County, 1937–1974.

to a cord and a half of pulpwood is needed to produce a ton of pulp, and a well-managed forest can produce about a cord per acre per year.

In 1978, one paper company, Georgia Kraft, owned 23,129 acres in Carroll County which were part of the company's total holdings of roughly a million acres in northern Georgia and Alabama. Georgia Kraft acquired most of its land in the late 1940s and early 1950s, when cotton was in decline and land prices were depressed. The company owns or leases properties of all size and shapes that are scattered over a radius of more than 100 miles. Despite their large landholdings, most pulp and paper companies depend on private landowners to produce a significant share of their raw materials. Georgia Kraft produces only about a quarter of its annual requirement of two million cords from its own land. The company buys pulpwood from small private landowners who do not have long-distance wood transport vehicles, through local collecting points, known as woodyards, which are scattered across the area at intervals of ten to fifteen miles (Fig. 7.4).

Changes in Land Use

Comparison of aerial photographs taken in 1937 with the 1974 land use map reveals the extent to which forest have taken over former agricultural land (Fig. 7.5). The land that has changed from agriculture to woodland does not lie in large solid blocks, but consists of small bits and pieces that are scattered all over the county. Carroll County, like the rest of the South, has a long tradition of turning livestock into the woods; pastures and even fields have shaded gradually into woodland, often without benefit of fences, and boundaries have not been as sharp or as clearly defined as in other parts of the United States. Farmers have had to exercise constant vigilance to keep brush and saplings from creeping up out of the ravines, first onto the steeper slopes that were less accessible to machinery, then onto the gentler upland slopes. Forest encroachment was rapid once that vigilance was relaxed. The process has been repeated on field after field all over the county, and probably throughout the Piedmont.

Concomitants of Change

Change does not occur in a vacuum, and one might have expected that Carroll County would have witnessed other changes while its farm land acreage has been declining. Any attempt to separate cause and effect in these changes would be of little value, and it is more reasonable to think of them merely as concomitants of land use change. Three of the more important concomitant changes have been the growth of employment in manufacturing, a massive redistribution of the rural population, and an increase in land speculation.

Employment in Manufacturing

Manufacturing began to develop in Carroll County at just about the time that farmers on forty to fifty acres discovered that they could no longer make money on cotton, and the growth of industry was able to absorb much of the displaced rural labor force. In 1930, almost two-thirds of all workers were employed in agriculture, forestry, and fisheries, but that figure dropped to less than one-thirtieth by 1970. Employment in manufacturing exceeded employment on the land for the first time in 1950, and by 1970 the number of persons employed in manufacturing was almost exactly the same as the number who had been employed in agriculture, forestry, and fisheries only forty years earlier.

The dominant manufacturing activity has been the production of textiles and apparel, which has a long tradition in the county. A textile mill with 500 spindles was operating at a dam on Snake Creek, near Whitesburg, as early as 1846, and a textile mill in Carrollton was the county's largest single employer in the 1920s.

Manufacturing employment has grown steadily in Carroll County since World War II, and the mix of industry has become a bit more diversified. By 1970, only half of the manufacturing labor force was employed in the traditional industries, and roughly a third had better-paying jobs in metal-working plants. In 1978, the Southwire Company, which was founded by a native son in 1950 to make specialized electrical wire and cable, had more than 3,500 workers on its payroll, and the county was agog with the news that it had been selected as the site for a new $50 million CBS Records plant that would employ 3,000 people. Nevertheless, five of the county's nine largest factories in 1978, and eleven of the sixteen that employed more than 100 workers, were still in the business of making textiles and apparel.

Migration to the Blacktop

The growth of industry provided factory jobs for rural people when they stopped cultivating farm land in Carroll County, and they had no need to migrate to other areas in search of work. Between 1880 and 1920 the population of the county grew consistently both in small towns (incorporated places) and in the countryside (unincorporated area), albeit at a slightly more rapid rate in the towns. After 1920, the population of the small towns continued to grow, but the rural population began four consecutive decades of steady decline as farm land was abandoned. This long-term downward trend was halted and turned around between 1960 and 1970 when the population grew slightly more rapidly in the countryside than in incorporated places.

Much of the growth in the rural parts of the county was the result of overspill from Carrollton and other incorporated places. The county shows few signs of any significant immigration from other areas, and its population "turnaround" seems to have resulted almost entirely from retention of its own natural increase. The county has been able to retain its young people because they can commute to jobs in the county or just across the county line; they no longer must migrate in search of work.

By 1970, the rural population had bounced back almost to its 1920 peak, but it had undergone a massive redistribution that might best be described as a "migration to the blacktop" (Fig. 7.6). Houses had been spread fairly evenly across the land in the 1920s, when the countryside was a place of work, but half a century later, when it had become primarily a place of residence, they were concentrated along the paved roads. Some of the people who lived along the highways had moved out from town, but many of them had merely moved up from the backcountry. The old folks had remained in the old family home until they died, but their children wanted a more modern place to live, and one from which it would be easy to commute on an all-weather road.

Rare is the stretch of blacktop highway in Carroll County that is more than a few hundred yards from the nearest house, and many of the houses along the highway are shiny new and modern. As soon as you get off onto the unpaved back roads, however, into the areas where farmland has been abandoned and allowed to grow up in brush, you begin to see unpainted, weatherbeaten wooden shacks with sagging, rust-blackened metal roofs– shacks that have long been unoccupied and are slowly tumbling down. All that remains of some is a

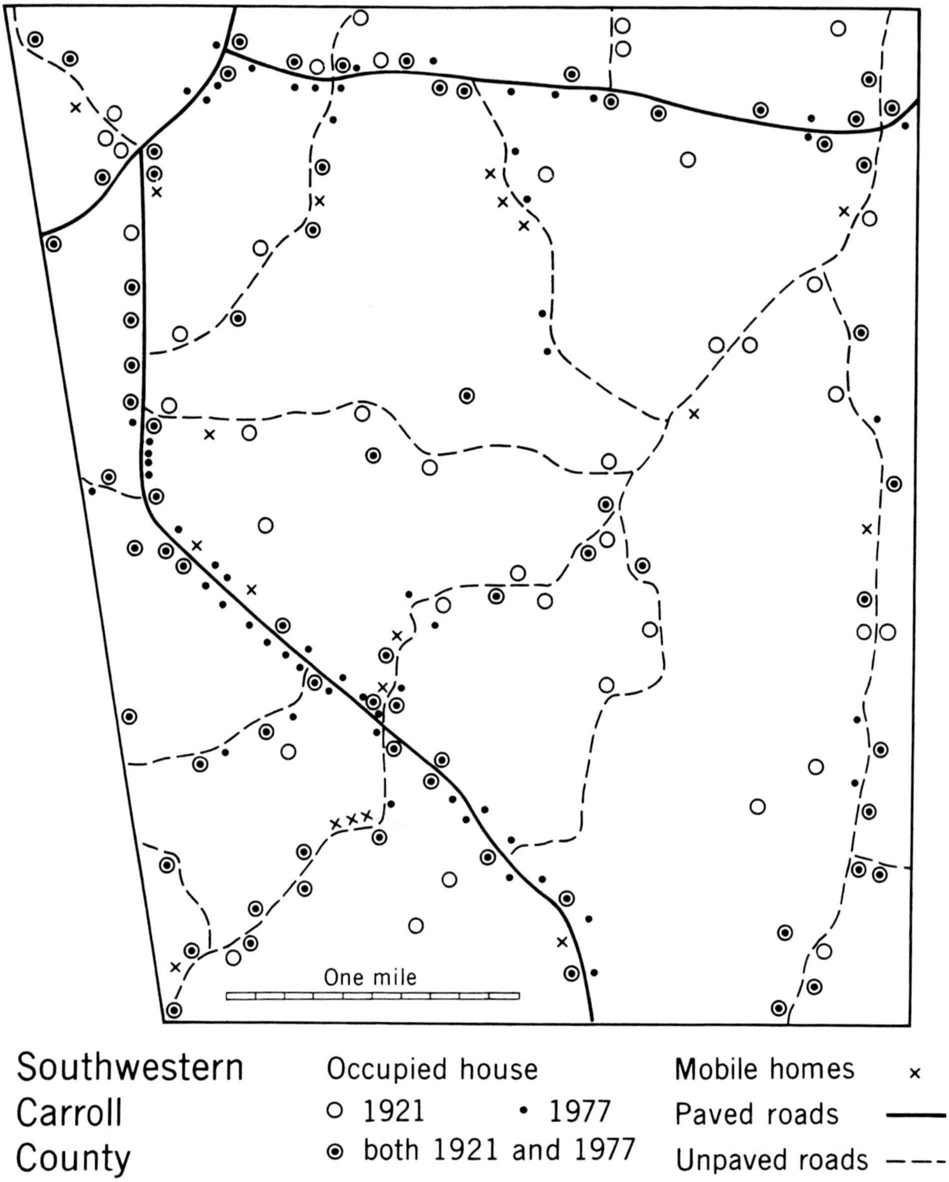

FIG. 7.6. "Migration to the blacktop," southwestern Carroll County, 1977.

lonely brick chimney or a forlorn patch of long-forgotten yard plants. The old house site is a good spot for a mobile home, however, because water is already available.

Strips of houses along the highway are a common, striking, and distinctive feature of much of the rural South. The houses normally line only one side of the road, and they rarely face each other. Otherwise, they could easily be in an urban subdivision, although they stretch for miles along rural roads at remarkable distances from the nearest place that could conceivably be considered urban. These highway strips do not seem to be as highly developed in Carroll County as they are in some other areas, yet they still are one of the more dramatic concomitants of land use change in the county. The land use map grossly underestimates their extent, however, because it does not show built-up areas smaller than the minimal mapping units of four hectares or narrower than 200 meters.

Speculation

I received predictably diverse answers when I asked different people about a "reasonable" price for land in Carroll County in the spring of 1979. Foresters told me that old, worn-out cotton land was worth about $150–$200 for forestry. Farmers were paying $600–$700 an acre for such land. Realtors told me that the price of five-acre lots in the northeastern half of the county—the side toward Atlanta—began at around $7,000, but lots of the same size were a bit cheaper in the southwestern half.

The difference in the style in which the price is quoted—price per acre versus total price for the property—probably is even more important than the considerable difference in prices. The forester and the farmer think in terms of cost per acre, but the city person looks at the total price tag. The farmer or forester values land in terms of what he can produce from it, and the more the better. He wants a large acreage, whose total price is too high for the city person, because he expects the land to pay for itself, and to provide him with a living. The city person values land in terms of amenity; he expects to pay for the land with money he earns elsewhere, and he wants a property at a price he can afford.

The small properties of Carroll County obviously have attracted buyers from urban areas in and near the county, because two-thirds of the land of the county is held by owners who are neither farmers nor paper companies. Some of these owners have built homes on their land and live on it, but some owners, especially those whose land is back from the blacktop, are doing little more with their land than paying taxes on it and holding it as an investment. Land, after all, is one of the few investments that has continued to hold its value in a period of frightening inflation, and it is available to the small investor if the individual parcel is small enough.

Policy Implications

Carroll County holds few surprises for anyone who knows the Piedmont, and therein lies its value because it is a useful model of land use changes in the region. In 1929, a quarter of the land in the county was used to grow cotton, but cotton ceased to be profitable in the 1930s, and the land was turned first to fodder crops, then to pasture, and eventually it was allowed to grow up in brush and trees. Most of the cleared land in the county in 1979 was used to pasture beef cattle. Three-quarters of the county's farm income in 1974 was derived from the sale of poultry, but modern poultry production is essentially a manufacturing operation that uses surplus farm labor to process raw materials imported from other areas. These imported raw materials are processed in small "factories" called poultry houses that happen to be on farms. A handful of entrepreneurs with exceptional managerial skills are still making a living from farming, but Carroll County has effectively ceased to be an agricultural area, and it should no longer be considered one when policy is formulated.

The forest area of the county nearly doubled between 1934 and 1974, and two-thirds of the county was wooded in 1974, but much of the woodland had limited economic value. Pine trees aggressively colonized abandoned farm land in the 1950s, and this pine forest has been a bonanza for pulp and paper companies, but it will not regenerate itself after it has been harvested unless it is carefully managed. The forest industries have acquired considerable acreages of land, and are managing it efficiently, but their land is inadequate to satisfy their raw material requirements, and they may encounter severe wood shortages unless some change in policy can encourage private woodlot owners to manage their forests more effectively.

Perhaps the needs of the forest industries are not paramount, and the land of Carroll County should be considered a long-term acreage reserve of potential cropland rather than a resource for the immediate production of forest products.

At the moment, the United States has more land than it needs to satisfy its domestic and export demands for food and fiber, but that situation could easily change and, in the future, the nation might need more cultivable land. Although it is not easy to think of abandoned farm land in such terms today, future Americans might see this land as a precious acreage reserve that can be brought back into production, land that has recouped much of its original fertility while it was "resting" under woodland cover.

The development both of agriculture and forestry in Carroll County has been constrained by the county's legacy of small landholdings, but the availability of small properties they can afford has encouraged speculation by successful business and professional people who have prospered while the county has made the critical transition from an agricultural to an industrial economy. The price of rural land in Carroll County, and probably much of the Piedmont, has been bid up by townspeople who are seeking a hedge against inflation. Of course, the land is precious to those who own it, and to those who would like to own it, but, to a truly astonishing degree, the land of Carroll County is in limbo.

Urban Encroachment

CHAPTER 8

on Rural Areas*

American cities have grown like Topsy. Their growth has been especially haphazard since the end of World War II, because the high-speed automobile and the high-speed highway have placed ever larger rural areas within easy commuting distance of urban jobs and because the jobs themselves have been moving from the center of the city toward its edges. The first city people to invade the countryside are those prosperous enough to purchase entire farms, where they can play at being farmers. These prosperous pioneers put ideas into the heads of real honest-to-goodness farmers, who begin to think of subdividing a strip of land along the county road, or some of the poorer land along the river or lakeshore, and selling lots to people. The farmer may sell his entire farm to a developer if he is close to the age of retirement, or has an undersized farm, or is farming agriculturally marginal land.

The developer needs a relatively large block of land, preferably near a fast road into the city. Piecemeal development of small lots near the city forces him to move ever farther outward in his search for a suitably large acreage. He often leapfrogs partially developed areas which already have city services and facilities to develop a large new subdivision in an area that lacks them. City people who desire new houses, especially inexpensive new houses, can find them in quantity only in such developments. Political reality dictates that the burgeoning population of these new developments must be given city services, even though existing capacity closer to the city remains unused. Taxes must be raised in order to provide the necessary services, and the farmers who remain are eventually

* "Urban Encroachment on Rural Areas," *Geographical Review* 66 (1976): 1–17.

forced to sell their land because they cannot afford to pay urban-level taxes.

Each of us is familiar with areas that have been paved over, built over, roofed over, and completely transformed from rural to urban during our own lifetimes. Many Americans, especially those who stick to the main highways, newspapers, popular magazines, and television, seem to believe that a tidal wave of asphalt and cinder block is moving inexorably outward from our towns, cities, and metropolises to inundate our countryside, leaving desolate islands of vacant or idle land among the developed subdivisions. An increasing national concern about the encroachment of cities on rural areas seems to be motivated by three more or less interrelated anxieties: (1) uncontrolled urban development is economically wasteful, whether in the cost of land, construction, and transportation or in the provision of such municipal services as streets, schools, parks, utilities, water, sewers, fire and police protection, garbage collection, snow plowing, and dog catching; (2) urban expansion is depriving the nation of some of its prime agricultural land; and (3) urban sprawl is an obvious aesthetic abomination that does irreparable damage to some rather vague but fearfully fragile thing that is defined only as "the environment" or, even more odiously, as "the ecology."

Proponents of controlled urban growth, to be sure, have run into strenuous opposition from some developers, who consider "sprawl" a vague, undefined, pejorative concept developed by elitists to condemn a particular kind of low-density, single-family growth that they happen to deplore. Despite what appears to be a widespread interest in the rate at which urban centers are expanding in the United States, many estimates of their growth, nay, even of their very extent, are only slightly more accurate and reliable than estimates of the incidence of adultery in this country. No federal agency is charged with the responsibility of collecting evidence on the extent of urban and built-up areas, and myths abound where facts are few.

Much conjecture about urban growth is strictly anecdotal. As an example, on September 3, 1975, Gladwin Hill wrote in the *New York Times*:

> "Open space is being lost to urbanization at the rate of more than 1,000 square miles a year—an area the size of New Jersey every decade. . . . In California, subdivision has reached a rate of 100,000 acres a year. A million acres of rural lots have been subdivided in Arizona . . . Once-bucolic Maryland typically faces the prospect of losing 45 square miles of priagricultural land to urbanization every year."

Attempts to estimate the national rate of urban expansion have varied widely. In 1956, Donald J. Bogue predicted that 371,800 acres of cropland and pastureland would be converted to urban uses each year between 1955 and 1975 in metropolitan counties alone. In the same year, D. A. Williams, administrator of the Soil Conservation Service, proclaimed that 1.13 million acres of cultivable land were being converted to urban use each year. This figure became doctrine in the Soil Conservation Service. In 1970, R. H. Lyddan, chief topographic engineer for the U. S. Geological Survey, blamed unidentified "land-use experts" for a mind-boggling estimate of more than 5 million acres a year. Three years later, J. Fulton Lutz of the North Carolina Agricultural Experiment Station came up with a national estimate of 2.92 million acres a year. In the same year, H. Thomas Frey of the U.S. Department of Agriculture provided an estimate of 740,000 acres a year, which was repeated in Newsweek (June 4, 1973). In 1974, Robert C. Otte, also of the U.S. Department of Agriculture, reported that the increase in urban land in Standard Metropolitan Statistical Areas (SMSAs) alone amounted to 709,600 acres a year. In 1975 [need more specific time/day], *Time* magazine gave a figure of 350,000 acres a year.

It is not always easy to track down the basis for such estimates. For example, on June 12, 1974, the *Wall Street Journal* reported that Congressman Morris Udall of Arizona had stated that "land equal in size to the New England states" would be developed in the next thirty years. A query to Congressman Udall elicited a letter dated June 24, 1974, from Dale Pontius, his staff assistant, which said:

> "To the best of my knowledge, Congressman Udall picked this up from Senator Jackson. . . . In a letter to the President on March 8th, 1974, Senator Jackson stated that between now and the end of the century an area the size of the states of New Hampshire, Vermont, Massachusetts, and Rhode Island. He went on to say that each decade an area greater than the size of the state of New Jersey would be absorbed by urban growth."

A letter to Senator Henry M. Jackson of Washington brought a response dated September 5, 1974, from Steven P. Quarles, special counsel to the Senate Committee on Interior and Insular Affairs, who said:

> Unfortunately, the land use policy files of the Committee . . . have

> already been placed in the archives. It would be very difficult at this point to trace the original citation for the estimate.

The Udall estimate works out at 1.554 million acres a year, the first Jackson estimate at 670,000 acres a year, and the second Jackson estimate at 481,000 acres a year.

Such a wild array of estimates, ranging from 350,000 to more than 5 million acres a year, indicates that some can be little more than sheer guesswork. Guesswork is not really necessary, however, because two sets of data provide independent evidence that can be used to estimate the amount of land which is being converted to urban use in the United States: data on "urbanized areas," published in each census of population since 1950; and data on the "urban and built-up" acreage of each county, compiled in 1958 and 1967 as part of the National Inventory of Soil and Water Conservation Needs. Both sources provide county-level data that can be mapped to identify the areas where cities are expanding most rapidly. The estimates of Frey and Otte were based on urbanized area data, but otherwise neither set appears to have received the attention it deserves.

Urbanized Areas

An "urbanized area" has been defined for each city of 50,000 persons or more at each decennial census of population since 1950. The urbanized area might properly be described as the "geographical city," because it includes the entire closely built-up area of the city and is enlarged as the city grows. Unlike the SMSA, which may include large tracts which exhibit urban characteristics in varying degree, the urbanized area is the real city, and it is redefined at each census. Urbanized areas are anathema to statisticians, because their size changes from census to census, but for that very reason they should be beloved of geographers, because they give a reasonably good idea of the extent and direction in which cities are growing.

In 1970, the United States had 248 urbanized areas which, together, covered just over 1.5 percent of the nation's total land area, although they contained fifty-eight percent of its total population. The total amount of land in urbanized areas increased from 12.8 million acres in 1950 to 24.98 million acres in 1960 and 35.08 million acres in 1970. The increases of 1.22 million acres a year between 1950 and 1960, and of 1.01 million acres a year between 1960 and 1970, might be misleading in two respects. First, the totals are not really comparable, because the original list of 157 urbanized areas in 1950 was increased to 213 in 1960 and to 248 in 1970. Second, even the total for 1970 cannot be considered a national estimate, because it is based only on the 248 largest urban centers in the United States. The omission of smaller centers may not be as serious as it might seem, however, because the twenty largest urbanized areas in 1970 contained 42.6 percent of the total acreage, whereas the twenty smallest contained only 1.4 percent.

The acreage of urbanized areas, not surprisingly, has a close statistical relationship with the size of their population; the coefficient of correlation between the 1970 area and the 1970 population of the 248 urbanized areas was an impressive 0.95. The residuals from this relationship, when mapped, appeared to have a completely random geographical distribution. Variations from one urbanized area to another appear to be most closely related to their size and age.

Three-fifths of the urbanized areas in 1970 had between 2,000 and 3,500 persons per square mile. This relatively narrow range of densities and the close relationship between population size and size of urbanized area suggest that population projections might be used to estimate future needs for urban land in the United States. A density of 2,000 persons per square mile is equivalent to 320 acres per thousand persons, and a density of 3,500 persons per square mile to 183 acres per thousand persons. Current population projections for the United States in the year 2000 range from 250 to 300 million people, an increase of fifty to 100 million between 1970 and 2000 [According to the U.S. Census, the total population of the United States in 2000 was 281,421,906]. An increase of fifty million people at a density of 3,500 persons per square mile would require only 9.15 million acres of new urban land; an increase of 100 million people at a density of 2,000 persons per square mile still would require only 32 million acres of land, appreciably less than the total urbanized area in 1970, and only 1.4 percent of the total land area of the nation.

The evidence of urbanized areas indicates that urban encroachment does not appear to represent a serious immediate threat to the nation's supply of rural land. Anyone who wishes to reject this conclusion has two valid grounds for doing so. First, the urbanized area data ignore urban growth at densities of fewer than 1,000 persons per square mile, which would gobble up land at a rate of 640 acres or more per thousand persons. Second, urbanized areas are not identified for places that have less than 50,000 persons, and encroachment at low densities from smaller places might represent a more serious threat to rural areas than high-density encroachment from our major cities. It is fortunate, therefore, that an independent source of information about urban encroachment on rural areas is available in the data collected as part of the National Inventory of Soil and Water Conservation Needs.

The Conservation Needs Inventory

The National Inventory of Soil and Water Conservation Needs, familiarly known as the Conservation Needs Inventory (CNI), represents the only attempt to collect comparable data for all major categories of land use in each county in the United States. The quinquennial census of agriculture, admirable though it is, collects information only on land in farms; the Forest Survey regularly collects information on wooded areas, but only on wooded areas; neither collect information on built-up and urban areas. The Conservation Needs Inventory collected data on all major uses of a two percent sample of the land in each county in 1958 and again in 1967. Local committees were instructed to identify, delimit, and exclude urban and built-up areas, all federal land (except cropland), and small water areas before the sample was drawn.

State totals for the Conservation Needs Inventory have been published in national summary reports, but the data for counties have been published only in separate reports for each state. The individual state reports can be obtained only by making direct application to the appropriate offices in each of the fifty states; the tediousness of assembling a complete set of the reports may be one reason why so little use has been made of their data.

The 1967 Conservation Needs Inventory reported that the total area of urban and built-up land in the United States was 60.993 million acres, or 2.7 percent of the nation's total land area. This figure is appreciably larger than the total of 35.08 million acres in urbanized areas in 1970, but presumably it includes the entire built-up area of the nation, not just the metropolitan centers. Nevertheless, most counties that have a high percentage of built-up area also contain metropolitan centers, and vice versa. The built-up area was highly concentrated. The 214 counties that were more than 12.5 percent built-up in 1967 occupied only three percent of the nation's total land area, but they contained 29.2 percent of its total built-up area and 52.2 percent of its total 1970 population.

Perhaps the most striking feature of the 1967 built-up areas is not how much land, but how little, has been urbanized. Large areas have no counties with as much as one built-up area in sixteen, and more than ninety-seven percent of the entire land area of the nation still awaited urban encroachment in 1967. Furthermore, the rate of encroachment between 1958 and 1967 should occasion no great alarm. Only 393 counties recorded an increase of as much as 1.8 percent in their built-up area between 1958 and 1967. Only 154 counties increased their built-up area at a rate of 0.5 percent a year or more, yet these counties accounted for nearly half of the total increase of 10.33 million acres between 1958 and 1967.

Unfortunately, the data on change in built-up area between 1958 and 1967 must be used with a certain amount of caution, because the data are not completely error-free. A number of county committees, especially in lightly urbanized counties, seem to have made no effort to update the 1958 figure for the 1967 inventory, because precisely the same figure is reported for both years. Some "changes" may have resulted from a consistent statewide bias in one of the two inventories. For example, the figures for built-up area in 1958 were suspiciously low for almost all counties in several states, notably Wyoming, Michigan, and North Carolina. Some of the increase reported in these states may be artifactual rather than actual. Conversely, counties in which the built-up area was overestimated in 1958 would show an improbably low increase, or even a loss. For example, it is not easy to believe that the built-up area of Cook County (Chicago, Illinois) dropped from 402.9 thousand acres in 1958 to 367.5 thousand acres in 1967.

Despite such flaws, the Conservation Needs Inventory data are the best available on the location and extent of built-up areas in all counties of the United States. Attempts to establish a statistical relationship between CNI data and the urbanized area data were only moderately successful; although both data sets are concerned with the same basic phenomenon, they measure slightly different aspects of it for slightly different units of area at slightly different times. The CNI measured the entire built-up area of an entire county in 1958 and 1967; the urbanized areas are concerned only with the built-up areas of single metropolitan centers in 1950, 1960, and 1970. Despite these differences, however, the two data sets reveal the same basic overall pattern: a high proportion of built-up area and a high rate of increase are associated with the larger metropolitan centers, and smaller places have had relatively little impact on surrounding rural areas.

Envoi

Urbanized area data from the census of population confirm the data collected as part of the National Inventory of Soil and Water Conservation Needs (CNI), which indicate that the built-up area of the United States in 1967 totaled 60.99 million acres and it was increasing at a rate of 1.148 million acres a year. If the current rate of growth continues, the total built-up area of the nation still would total only 98.88 million acres by the year 2000. An even smaller estimate is produced if current urban population densities of 180 to 320 acres of land per thousand persons are used to project needs for new urban land; an increase of 100 million persons at the lower density would require only 32 million acres of new urban land by the end of the century. The total land area of the United States is 2,271,343 thousand [2.27 billion?] acres.

It seems reasonable to conclude that little more than four percent of the nation's land area will be urbanized by the year 2000 and that urban encroachment will not remove significant acreages of land from agricultural production within the foreseeable future. Most of the new urban land will be in and near major metropolitan areas where most of the built-up area is already concentrated.

Half a Century

CHAPTER 9

# of Cropland Change*

You and I both know that suburban development is encroaching on productive agricultural land in the United States, because we have the evidence of our very own eyes. Each of us can think of former fields that have been converted to nonagricultural use. But how representative are our observations? Advocates of growth control argue that this country is losing farmland at an alarming and unacceptable rate. Should we share their concern?

We know that the use of cropland changes from year to year and that, in any given year, a significant fraction of our cropland, perhaps one-fifth to one-third, does not actually produce any crops at all. Drought, flood, storm, diseases, insects, and other destructive forces of nature can nullify the best-laid plans of farmers, but complete crop failure is only part, and usually quite a small part, of the explanation. Good farmers rotate their crops, and many rotations require that in some year of the rotation each piece of ground lie fallow, or be used only for pasture, or grow only cover crops that are not harvested.

The inherent capability of soil and climate to produce crops is only one of many factors that influence the way farmers use cropland. Land may also lie uncultivated temporarily for idiosyncratic, nonagronomic reasons: family squabbles over inheritance or divorce, compliance with or circumvention of government farm programs, quirks of wealthy landowners who do not have to extract income from every acre they possess. The use of cropland is constantly changing, and changes from year to year or from census to census may be only short-term fluctuations that are little more than statistical "noise."

---

* "Half A Century of Cropland Change," *Geographical Review* 91 (2001): 525–43.

A clear picture of cropland change in the United States requires an analysis of long-term trends that places the distractions of short-term noise in proper perspective. The dawn of a new century seems an appropriate time for a close examination of the geography of cropland change in the preceding half-century. Fifty years is time enough to provide a balanced view of long-term trends and patterns and a more robust and defensible base for judging whether suburban encroachment and development "threaten the future of our agricultural land base" (American Farmland Trust, 1997, xi).

The National Agricultural Lands Study

The first major false alarm about "our vanishing farmland" was sounded in 1979, when the National Agricultural Lands Study (NALS) made scare headlines by asking, "Will there be Florida oranges, fresh milk from St. Johnsbury—and Michigan tart cherries? Florida—producer of more than half of the world's grapefruit and one-fourth of its oranges—will lose virtually all of its unique and prime farm land by the turn of the century if present land loss trends continue" (Fields, 1979, 14).

Serious students of land use immediately recognized that the NALS estimates were grossly exaggerated (Fischel 1982; Raup 1982). The NALS had derived these estimates from a comparison of two sample surveys made by the U.S. Soil Conservation Service in 1967 and 1975. The comparison was made even though the 1975 survey was conducted in great haste, had different objectives, used different procedures and a far skimpier sample, and was never intended to be comparable with the 1967 survey. In 1984, the U.S. Department of Agriculture essentially disavowed the NALS estimations, but these inflated estimates still circulate, and some people persist in citing them (American Farmland Trust, 1997, xi).

Florida illustrates the errors that an undersized sample can produce. James R. Anderson, head of the Office of Geographic Research at the U.S. Geological Survey, was so concerned about the errors in the NALS data that he dispatched Richard Kleckner to the SCS office at the Iowa State University Statistics Laboratory to compare the NALS estimates with other data on a county-by-county basis in an attempt to find the source of error.

Kleckner discovered a major clerical blunder. A clerk in one Florida county had erroneously listed 6,500 acres of urban land in the Rural column and

367,850 acres of rural land in the Urban column (Kleckner 1980). This error was magnified sixty-seven times when the statisticians expanded the county sample data to totals for the state and the nation. In hindsight, which of course is always perfect, it turns out that Florida actually had 44,000 more acres of citrus groves in 1997 than in 1982, even though the great Christmas freeze of 1983 devastated 200,000 acres of grove land that has not been returned to citrus production.

The alarming forecasts of the NALS may have contributed to the "farm crisis" of the early 1980s, which was actually a crisis for nonfarm investors. These forecasts were widely publicized, and they encouraged nonfarm investors to buy farmland at greatly inflated prices, because they thought it was becoming scarce.

Most farmers tightened their belts and lived on their capital when the price of farmland crashed nationwide in the early 1980s, but nonfarm investors lost their shirts, and their squeals of pain were widely publicized. In retrospect, the "crisis" was little more than a blip in long-range trends in the declining number of farms and the slowly escalating price of farmland.

Discordant Data Sets

Students of cropland change can be thoroughly confused by the disarrayed U.S. Department of Agriculture, which publishes two different sets of "official federal statistics" that do not agree. Both the census of agriculture and the National Resources Inventory (NRI) publish data for the years ending in -2 and -7, but there the similarity ends. For example, the census says that the nation gained 13.64 million acres of harvested cropland between 1992 and 1997, but the NI says that it lost 8.46 million acres of cultivated cropland.

A discrepancy almost as large as the state of Indiana cannot be dismissed as trivial, even though it is equivalent to little more than one percent of the total land area of the conterminous United States. This discrepancy is significant in its own right, and it has fueled disputes about the significance of farmland loss in the United States, because disputants quite naturally have elected to use the "official" data that best support their positions. Why do the data disagree?

The National Resources Inventory and the Census of Agriculture

The NRI collects data at "scientifically selected sample sites," and uses "sophisticated statistical procedures" to estimate national and state totals of land-use acreages (NRCS, 2000). The Iowa State University Statistical Laboratory collaborated with the NRI in designing a permanent network of 300,000 sample sites. Repeated use of the same sites facilitates recognition of change. Most sites are 160-acre squares. Actual field visits to sites rarely are necessary, because photointerpreters identify land uses and land-use changes on aerial photographs.

The NRI uses sampling rates that vary from two to six percent. It has made a herculean effort to project data from small samples to national estimates, even though it is aware that an error at any sample site will be multiplied twenty to fifty times when it reaches the national total. This process, like all statistical procedures, has a predictable margin of error, which generally becomes larger as numbers and areas become smaller.

National NRI estimates, for example, are more reliable than are state estimates, and the NRI does not publish data for areas smaller than states. The margin of error for the estimate of cultivated cropland at the national level in 1997 was a respectable 0.6 percent, but at the state level it varied from 2.2 percent in Texas to 67.4 percent in Rhode Island. The margin of error for the estimate of change in cultivated cropland between 1992 and 1997 was even more volatile; it was 8.8 percent at the national level, and in fifteen states the margin of error actually was greater than the estimated change.

Despite these problems, the NRI appeals to some users because it estimates the amount of land converted from one use to another—for instance, from cropland to developed land—at the national level. It also has useful cross-tabulations at national and state levels of land use by soil erosion, prime farmland, and various aspects of wildlife habitat.

The NRI has a proclivity for publicity. It disseminates its results aggressively to the news media, which can be embarrassing when it makes a mistake. For example, a programming error in the statistical software necessitated a revision of the 1997 NRI. Nevertheless, the media cite the highly publicized NRI data more often than the more reliable data of the staid old census of agriculture, which assumes that its results speak for themselves and require no advertising.

The quinquennial census of agriculture is a complete enumeration of all farm operators in the United States, and it is our only source of consistent, comparable (both geographically and historically) agricultural data at the county, state, and national levels. The census sends each farm operator a detailed questionnaire that makes the much-maligned long form of the census of population look like child's play, and it publishes the tabulated results in massive volumes for each state. The census of agriculture defines its terms precisely, it scrupulously reports any changes in the definitions it uses, and it clearly identifies any data that are not fully comparable with the data in earlier censuses.

Farmland or Cropland?

The people who lament the loss of farmland and open space cite often fail to identify the source of their estimates. They use terms such as "rural land," farmland," "agricultural land," "cropland," "fertile land," "open space," and "countryside" as though they were interchangeable, although they are far from synonymous, and only "farmland" and "cropland" are defined precisely enough at the national level to have any analytical value.

One might suppose that, at the very least, we ought to know how much *rural* land we have in this country, but we do not. Since the 1930s, aerial photography has given us the technology we need to prepare an accurate and reliable inventory of land use in the United States, but it would be so expensive and time-consuming that money to pay for it has rarely been available. The only attempt to compile a complete national inventory of land use was the land-use and land-cover mapping program that Arch Gerlach and James Anderson spearheaded at the U.S. Geological Survey between 1974 and 1982. That program was a one-shot effort that has never been repeated. We remain woefully ignorant about where and how much our built-up area is encroaching on rural land.

Even the official census definition of the rural population is backhanded and negative: the rural population includes everyone who is not urban! One might estimate the area of rural land in the United States by adding up the total acreage of urban areas and subtracting this total from the total acreage of the nation, but that would be an exercise in futility, because the census defines most urban areas by their political boundaries, which are extremely tough to change, so the urban area of the United States, as officially defined, hardly varies from census to census.

"Farmland" is defined precisely enough to be useful analytically, but it is flawed conceptually as a measure of agricultural land use and land cover. The census of agriculture defines farmland as all land that was part of a farm

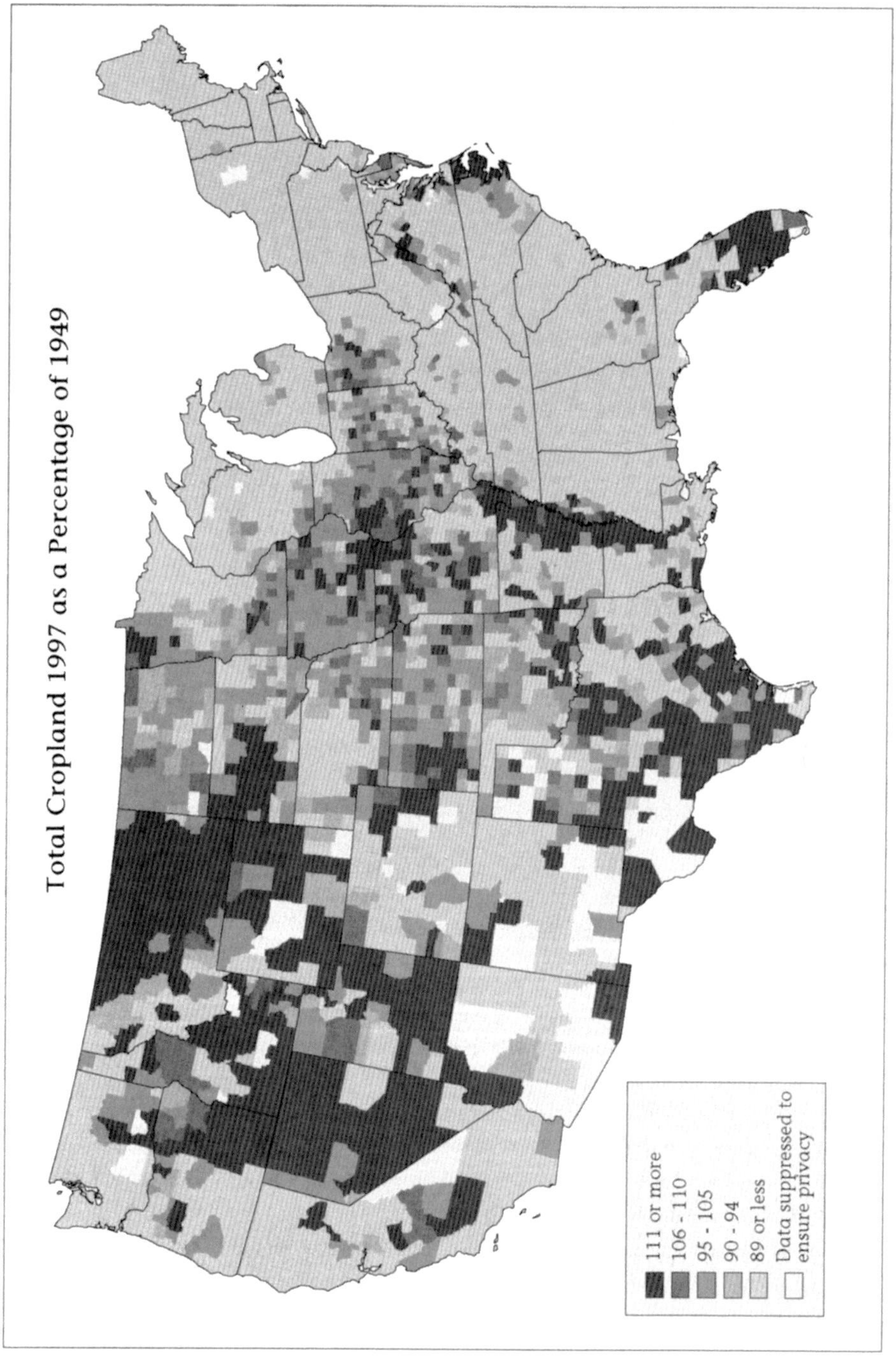

FIG. 9.1. Total cropland in 1997 as a percentage of 1949.

operator's total operation, and it has always defined farm operators quite permissively; since 1974, anyone who has produced $1,000 or more of farm products during the census year has been classified as a farmer. The farmland area of the United States peaked in 1949, at a total of 1,161 million acres, and it had declined to 932 million acres by 1997.

This change in the acreage of farmland is a misleading indicator of agricultural changes, however, because part of the land owned by farmers is woodland, which contributes scantily, if at all, to the farm economy, and this land immediately ceases to be classified as farmland if the farmer sells it to a lumber company, to a pulp and paper company, or to any other nonfarm entity. Although the United States "lost" 229 million acres of farmland between 1949 and 1997, 149 million acres (sixty-five percent) of this land was woodland that farmers sold to nonfarm entities, and its use and appearance may not have changed at all.

The acreage of land that farmers actually use to grow crops would seem to be the best measure of effective farmland. The census of agriculture provides data on the total acreage of cropland in each county and on the acreage that farmers actually managed to harvest in the census year. Total cropland includes all land from which crops were or could have been harvested in the census year.

The state of California illustrates the kind of data that are published in the census of agriculture for the entire nation, for each state, and for each county. Between 1949 and 1997, the state lost nine million acres of farmland, but two-thirds of the lost farmland was woodland that farmers had sold to nonfarm owners. The state lost three million acres of land that was classified as cropland, but much of this land did not produce crops in 1949, and the acreage of cropland harvested in the state actually increased by more than half a million acres.

Percentage Change of Cropland

I used the total acreage of cropland in each county in the United States as the best measure of effective farmland, and I explored how the geography of total cropland changed in the latter half of the twentieth century. The total cropland area of the United States dropped from 478 million acres in 1949 to 431 million acres in 1997. The decline has been reasonably steady, with no abrupt changes, no evidence of any "shift in trend" (Fig. 9.1).

A net loss of forty-seven million acres in forty-eight years works out to nearly one million acres a year. Is the loss of ten percent of our cropland in half a century, or two-tenths of one percent a year, cause for concern? At this rate, the half-life of our remaining 431 million acres of cropland is about 220 years, or somewhere around the year 2220, and even that projection assumes that none of the cropland that has been "lost" will ever be brought back into agricultural production.

The map of percentage change of cropland between 1949 and 1997 is complex, and its close interpretation requires intimate local knowledge, but some broad regional patterns do stand out. In the Midwest, most counties changed only slightly. The West is a crazy quilt of large counties with small acreages of cropland and extreme changes. In the East, South, and Upper Great Lakes area, virtually every county lost more than ten percent of its cropland. Cities are difficult to find on all parts of the map, because many counties with large cities lost only a few percentage points more cropland than did the other counties in their area.

The most striking feature of the map is the lack of change in the Midwest, the nation's agricultural heartland. A ragged-edge wedge of stable or slightly increased cropland narrows eastward from the Great Plains toward Columbus, Ohio. Its northern edge runs from Minneapolis to Chicago to Cleveland, and its southern edge runs from Columbus to Oklahoma City. The acreage of cropland in most of the counties in this area changed less than ten percent, some even less than five percent, between 1949 and 1997.

For example, the acreage of cropland in the state of Iowa, the very heart of the heartland, actually increased three percent. The state added 772,525 acres of cropland, an area approximately the size of Rhode Island, between 1949 (26,049,319 acres) and 1997 (26,821,444 acres). Iowa was far from alone, but after all of the hullabaloo about our vanishing farmlands, some people may be surprised to learn that the acreage of cropland actually increased more than ten percent between 1949 and 1997 in one of every seven counties in the United States (and, with all due respect to the NALS, one of every four in Florida!).

The largest blocks of counties where the cropland acreage has increased ten percent or more are where farmers have extended irrigation works in the intermountain West, plowed up the range on the northern Great Plains, dug deep wells to tap the Ogallala aquifer on the southern Plains, cleared and drained the bottomland hardwood forests of the Delta (the alluvial plain of the lower Mississippi River), and drained swamps in eastern North Carolina and southern Florida, where citrus production has expanded into areas that were considered unsuitable for citrus before the great Christmas freeze of 1983.

Critics may protest that most of these increases are one-shot affairs, that they may not be sustainable, and that they are not likely to be replicated in the future, but few of them could have been foreseen in 1949, and it is not unreasonable to assume that technological innovations may be responsible for similar changes in the next half century. For example, the eradication of boll weevils has already enabled farmers in Georgia to bring more than one million acres of land back into cotton production, and 158,000 acres of hillside oak woodland in Sonoma County, California, which are marginal farming because of their slope, drainage, soil, or other qualities, are ripe for conversion into vineyards when the price is right. Such a conversion would more than double the county's 1997 cropland acreage.

Absolute Change of Cropland

A map of absolute change (Fig. 9.2) must complement a map of percentage change because percentage change alone can be misleading, especially in the West, where many counties are large but have such limited acreages of cropland that a small absolute change can produce a huge percentage change. Percentage change is influenced by the size of the base on which it is calculated, and some of the counties with the largest percentage increases in cropland between 1949 and 1997 were those that started off with small acreages of cropland in 1949; a large percentage of very little is still not very much.

The gain or loss of a few thousand acres of cropland would be a mere drop in the bucket in most Iowa counties, but it could produce a major percentage change in many counties in the West, with impressive, perhaps even misleading, cartographic consequences. For example, Brewster County, in the Big Bend country of southwest Texas, shot from 914 acres of cropland in 1949 to 5,682 acres in 1997, for a stunning increase of 522 percent, and Brewster County takes up more space on the map than do the states of Connecticut and Rhode Island combined.

The dot map of cropland gain shows that slight absolute increases were responsible for the large percentage increases in the intermountain West, but it also highlights many other areas of gain: the Palouse, Snake River Valley, San Joaquin Valley, several areas on the Great Plains and in southern Texas, the

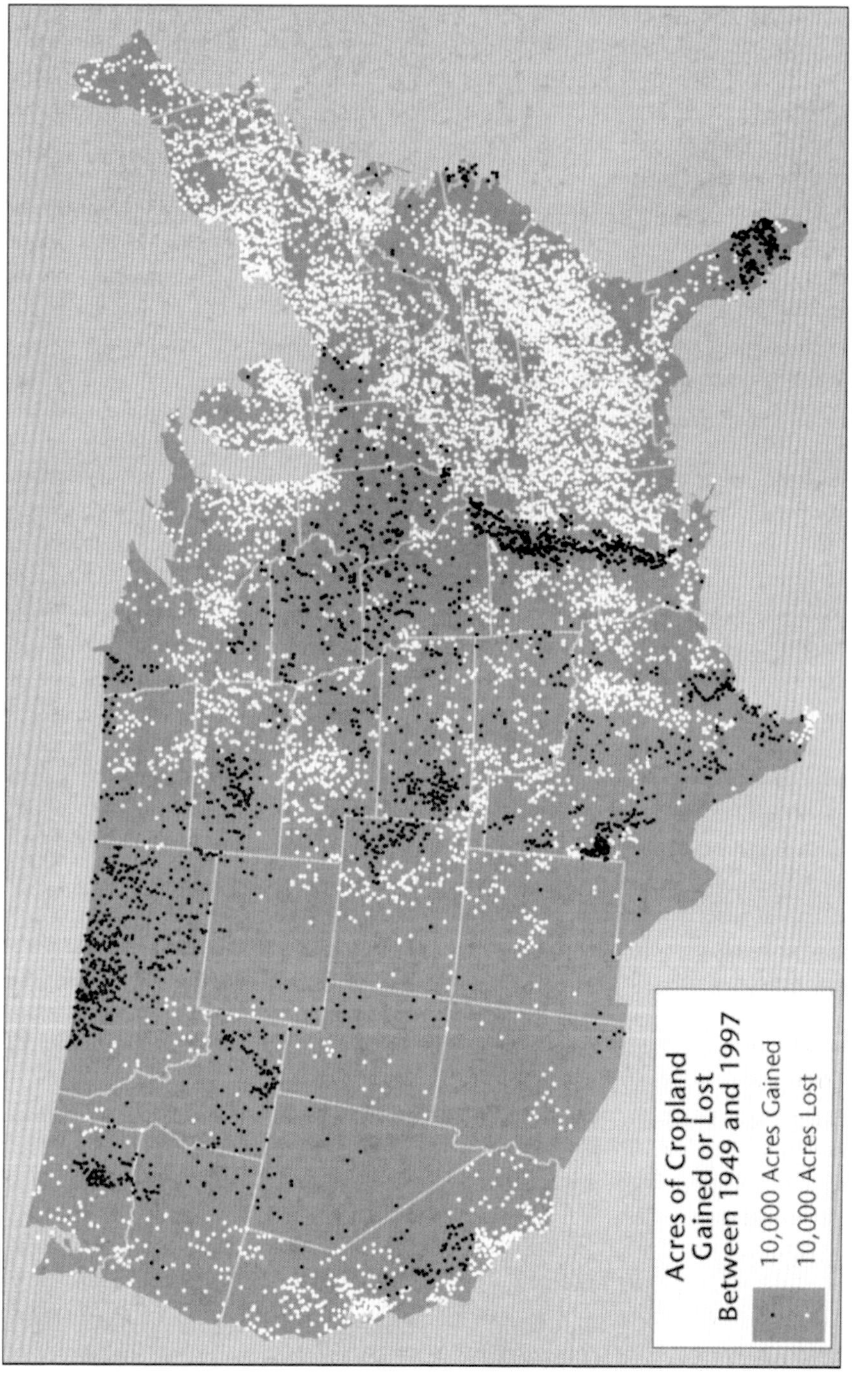

FIG. 9.2. Acres of cropland gained or lost between 1949 and 1997.

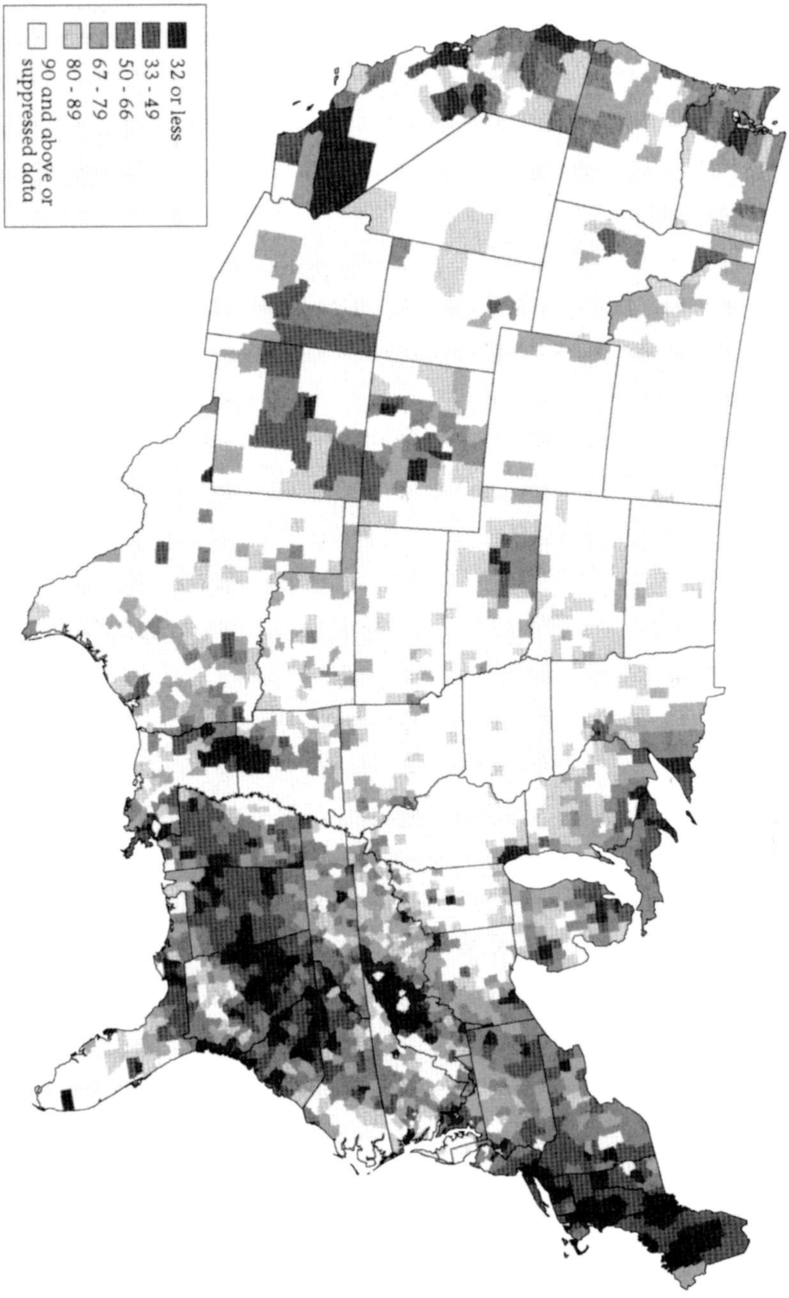

FIG. 9.3. Heavy losses: total cropland in 1997 as a percentage of 1949.

Delta, eastern North Carolina, and southern Florida. The dot map also shows the ubiquity of incremental cropland increase in the agricultural heartland, where the drainage of small pockets of wetland added a few thousand acres to the cropland base of many counties.

The most striking feature of the dot map of cropland loss is the ubiquity of loss in the entire eastern half of the nation, except for the Midwest and the Delta. Some major cities are evident, but finding many is a real challenge, and it is obvious that urban encroachment removed far less land from cultivation than did Old Mother Nature. The areas with the greatest loss of cropland are marginal agriculturally, with soils of low inherent fertility and topography unsuited to modern farm machinery. Much of this land probably never should have been cultivated in the first place.

Los Angeles and San Francisco stand out, but Portland and Seattle are difficult to find. You can spot Minneapolis/Saint Paul, Chicago/Milwaukee, and Detroit, but you cannot invoke urban encroachment to explain the heavy loss of cropland in the Nebraska Sand Hills. Kansas City and Saint Louis are evident, and urban encroachment may explain the loss of cropland on the Blackland Prairie of Texas, but not on the sandy lands of eastern Texas, Arkansas, and Louisiana. You can find Cincinnati and Columbus, but, east of Columbus and south of the Ohio River, cities are tough to distinguish from the marginal agricultural counties around them.

I made a map with more detailed resolution of the counties that lost more than ten percent of their cropland in an attempt to identify possible urban influence, and once again it requires close and careful study (Figure 9.3). You can find signs of urban impact, especially on the West Coast and around the edges of the Midwest, but the largest areas of greatest cropland loss were in the hills of the Northeast and Appalachia, on the tired, old, worn-out soils of the Piedmont, and on the thirsty sands of the Coastal Plain.

Extensive areas in the South have lost large areas of cropland because of stagnant demand and loss of market for cotton, tobacco, peanuts, and other traditional crops, and dramatic increases in yields have substantially reduced the acreage of cropland necessary to produce the same amounts of these crops. Many of the counties that have lost cropland would dearly love to have some urban encroachment, or any other kind of economic development, for that matter.

Some critics draw on the experience of medieval Europe, where only productive farmland could support cities back in the days of oxcarts. They claim that the cropland we are losing near our cities is prime and unique, our most valuable farmland, but they are confusing price with value. Cropland at the city's edge always has the highest price tag because it is ripe and easy for conversion, but its price is based on its geography, its location at the bow wave of the expanding city, not on its inherent productivity, and when it is converted to urban use the high price tag simply slips to the next land farther out, regardless of its quality.

Is California Different?

Californians protest that their state is different, because urban areas are encroaching on critical farmland that produces the specialty crops that contribute diversity to the American food mix. Certainly, the state leads the nation in lamentations about the loss of valuable farmland, but the census of agriculture says that its orchard and vineyard area actually increased by more than 1.1 million acres between 1949 and 1997 and that the area producing vegetables increased by more than 670,000 acres. The acreage of high-value-added field crops (cotton, corn, rice) increased, while the acreage of lesser-value small grains (barley, oats) declined. Many of the orchards, vineyards, and vegetable fields of 1949 have been paved over, to be sure, but new orchards, vineyards, and vegetable fields have been developed on land that once produced crops of lesser value, or no crops at all.

It seems that the specialty crops produced in California can be grown—and, indeed, are grown—in other parts of the state and in other parts of the United States. The conversion of specialty cropland to urban uses is highly visible, because it occurs near the cities where the most people live and can see it. Far less visible are the farmers who have sold their land in these areas. They have taken their windfall profits and quietly invested them in replacement operations elsewhere that produce the same products.

Epilogue

Rates of cropland loss in the United States during the second half of the twentieth century suggest no reason for immediate national concern. Even those counties that lost as much as twenty percent of their cropland in the half-century have enough left to last for two more centuries at that rate, and most of the counties in the agricultural heartland in the Midwest can continue on forever at their current rates. One cannot justly invoke the loss of cropland at the national scale as an argument for blocking suburban development in any local area.

The reaction to this conclusion ranges from agreement to anger to incandescent outrage. Why do so many people seem to feel so threatened by the idea that the United States is not losing cropland at a significant rate? Part of the answer is that each of us knows, from firsthand observation, that our cities are gobbling up our agricultural land. Each one of us can think of rural areas that we have seen converted to urban uses. We yield to the perfectly normal human temptation to assume that our observations and experiences are typical, and we project them to the status of Universal Truths. We are predisposed to believe the worst. The American news media and the American people have an apparently insatiable appetite for news of disaster.

The best evidence that the United States is not losing significant amounts of cropland is our persistent national "farm problem": overproduction. Rarely has anyone ever asked American farmers to produce more of anything, and they are often urged to produce less. They can produce entirely too much of any midlatitude product you care to mention. In fact, the United States has so much good cropland that each year we export one-third of it, which is a shorthand way of saying that we export one-third of the crops we produce, and nobody complains that we are exporting too much. Quite the contrary. Many people are trying to figure out how to increase our agricultural exports, and we routinely suffer farm crises when our export markets collapse.

Even California has too much good farmland: The state exports sixty-eight percent of its agricultural production (fifty-five percent to other states) (Medvitz 1998). Furthermore, some economists would argue that good agricultural land resists urban encroachment better than does unproductive land and that, if farmland were truly scarce, the invisible hand of the market would push its price up high enough to protect it (Brueckner, 2000). Perhaps the price of the vineyard land in the Napa Valley is a case in point.

The loss of cropland in the United States is primarily a function of land quality rather than of urban encroachment. The acreage of cropland has declined the most in areas where the economics are toughest, where the land is not worth farming at the present time. Only a modest fraction, albeit a highly visible fraction, of our cropland has been lost to suburban development, primarily in areas close to our largest metropolitan areas. There are good and compelling reasons for concern about the expansion of urban areas into the adjacent countryside, but attempts to block or control it cannot be justified on the grounds of loss of good cropland/farmland.

Resort Areas

CHAPTER 10

in Wisconsin*

Visitors in Wisconsin often are surprised when they hear people talk about going to "the lake," as though the state had only one, rather than 14,000 or more. In a sense, however, for many residents the state really does have only a single lake, the one that is graced by the family's second home. A summer cottage on a lake has been the very symbol of the good life for several generations in Wisconsin and in Michigan and Minnesota as well. Surrounded by whispering pines, the cottage symbolizes relief from the heat of summer pavements; it symbolizes relaxation and escape from the daily routine of the office or shop; and, most of all, it symbolizes having arrived. The ability to afford a second home is an important status symbol in a society that equates worth with income. Of course, income may not be flaunted; it must be manifest in subtle ways, such as living in the right suburb, driving the right car, wearing the latest fashions, having an office in the proper area, sporting a midwinter tan, and having a place on the lake, preferably on the right lake, because some lakes are "better" than others.

The lakeshore cottage areas of the Upper Great Lake states (Minnesota, Wisconsin, and Michigan) attracted the attention of researchers and the popular press during the late 1970s and early 1980s, when it became apparent that population growth had begun to slow down and even decline in many major metropolitan areas, whereas the population of many nonmetropolitan areas was continuing to grow. The catch phrases "population turnaround" and "rural renaissance" were coined to describe the fact that the total aggregate

* "Resort Areas in Wisconsin," *Geographical Review* 74 (1984): 192–217.

population of nonmetropolitan counties was growing at a faster rate than the total aggregate population of metropolitan counties. Much of the population growth in the nonmetropolitan counties of the Upper Great Lake states during the 1970s patently was the product of overspill from metropolitan centers and smaller towns and cities. Many counties whose growth could not be explained by urban overspill were in well-known lakeshore resort areas of the Great North Woods, areas that have significant concentrations of second homes and summer cottages. Some researchers have assumed that the population of these counties has grown as a result of the inmigration of elderly people who have converted their lakeshore cottages into their primary homes and permanent retirement residences.

The "natural history" of summer cottages in the Upper Great Lake states is a familiar story. Many began as primitive shacks with minimal creature comforts, mere "hunting lodges" where men could retreat to hunt, fish, play poker, smoke cigars, drink whiskey, and enjoy the great outdoors. When womenfolk decided that they had a right to enjoy the great outdoors and began to frequent the cottages, they almost immediately demanded the comforts of civilization, such as running water, electricity, and telephone. As the owners aged, they realized that the cottage would be a retirement place, and they winterized it into a year-round residence for that purpose. If it was close enough to a city, usually fifty or so miles, it might even be sold to a long-distance commuter for a permanent home.

The resort communities have changed as much as the individual cottages. In the early days, the range of possible activities was restricted chiefly to hunting and fishing, or just sitting, rocking, and thinking. Access has become much easier since World War II, and a steady increase in the influx of tourists has emboldened local people to develop more comfortable accommodations, a greater variety of activities, and vastly improved services for visitors. Jack Ruttger's Bay Lodge, north of Brainerd, Minnesota, shows how resorts have changed. His grandparents let people sleep in the hay mow of their barn. Then they built a cottage and added more, until the resort now has sixty units that can accommodate 200 people, with condominiums for sale. The lodge has indoor and outdoor swimming pools, a whirlpool, and saunas, in addition to the swimming beach on the lake. It offers canoeing, sailing, water-skiing, golf, tennis, horseback riding, special activities for children, bingo, movies, and bonfires on the beach.

The tourist season, which once lasted only from the Fourth of July until Labor Day, has gradually been extended into the fall and winter. Some visitors prefer the quieter period after children have returned to school, some like to come in fall when the leaves are changing color, and many are enticed by a whole new gamut of winter activities that includes snowmobiling, cross-country skiing, and ice fishing. The extended tourist season has created labor shortages for the operators of some summer businesses, who could employ cheap student workers when the season meshed with the academic calendar. Tourism interests in Wisconsin have even introduced bills in the state legislature that would delay the opening date of schools and colleges.

The extended season has created more and better jobs for local residents, who no longer must earn their entire annual income in the eight short but hectic weeks between the Fourth of July and Labor Day. They are employed for a longer period, because the tourist establishments remain open for a greater part of the year. Perhaps even more important is a multiplier effect that does not really begin to operate until a resort area has attained the crucial threshold size necessary to provide satisfactory employment for large numbers of local persons for a significant part of the year.

Past this threshold, for example, there is enough employment in construction to keep local workers busy almost year-round in winterizing old summer cottages and in building cottages, condominiums, lodging and eating places, stores, and other business establishments. Furthermore, most resort areas are rural, and few rural areas have the municipal services that city folk take for granted, so employment is available almost year-round in well drilling, construction of septic systems and sewage-holding tanks, garbage removal, snow plowing, yard maintenance, domestic service, and caretaking.

Young people in the resort areas can now find jobs near home, and they no longer have to migrate to the metropolis in search of work when they graduate from high school or complete a tour of military service. The employment of local young people in the tourist industry and related activities has greatly reduced out-migration, which formerly assumed massive proportions, and the retention of young people may have been as important as the in-migration of retired people in explaining the growth of population in some resort areas.

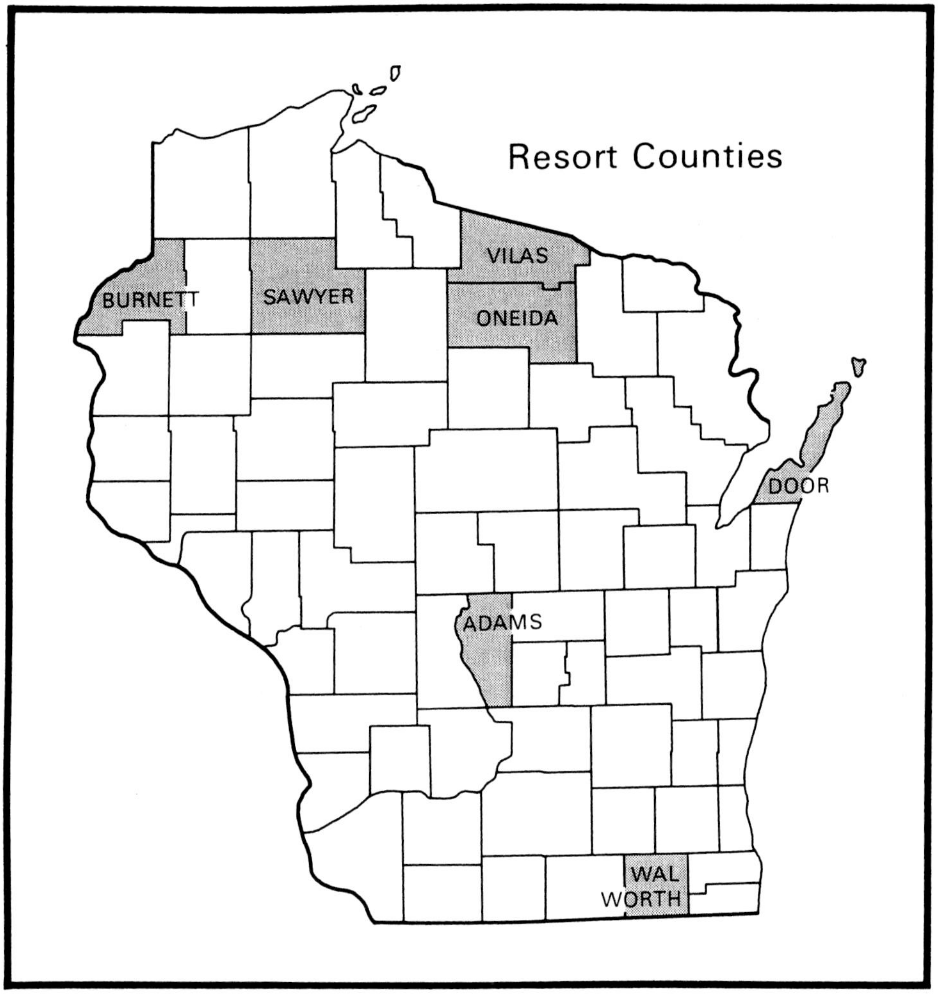

FIG. 10.1. Well-known resort counties in Wisconsin.

Identification of Resort Areas

In this essay, I examine three distinctive resort areas in Wisconsin. Resort areas must be well known and widely publicized if they are to attract visitors, and "almost everyone" in Wisconsin knows that Burnett County, Sawyer County, Vilas County, Oneida County, Door County, Adams County, and Walworth County are popular resort areas (Fig. 10.1). Common knowledge is not always correct, however, and I used three sets of data to verify resort areas in Wisconsin: (1) the number of seasonal vacant housing units in each county; (2) the number of persons per housing unit in each township; and (3) the number and gross sales of all hotels, motels, and resorts in each county.

Seasonal Vacant Housing Units

The U.S. Census of Housing has published data on the number of seasonal vacant housing units in each county since 1940. This category includes beach cottages and hunting cabins, units used for summer or winter sports, units held for herders or loggers, and units held for migratory farm workers. I took the number of seasonal vacant housing units in each county in 1940 as a reasonable surrogate for the number of summer cottages in a county.

In 1940, summer cottages were concentrated near the major cities (Fig. 10.2). One-quarter of the cottages in Michigan were within fifty miles of Detroit, two-fifths of the cottages in Wisconsin were within fifty miles of Milwaukee, and almost one-third of the cottages in Minnesota were within thirty miles of Minneapolis. Counties in the lake areas of southwestern Michigan and in the dune areas along the western side of the state had concentrations of cottages, and so did half a dozen counties in the northern part of the Lower Peninsula. The summer-cottage belt across northern Wisconsin and northern Minnesota was still embryonic in 1940, however, and only Vilas and Door counties in Wisconsin had more than two cottages per square mile.

In the 1980 Census of Housing, some units that would have been classified as seasonal vacant in 1940 were placed in a new category: held for occasional use. This category includes units held for weekend use, plus shared ownership and time-sharing condominiums. It also includes most second homes, although the definitional confusion was further roiled by the candid admission that "some second homes may be classified as 'seasonal'" rather than as "held for occasional

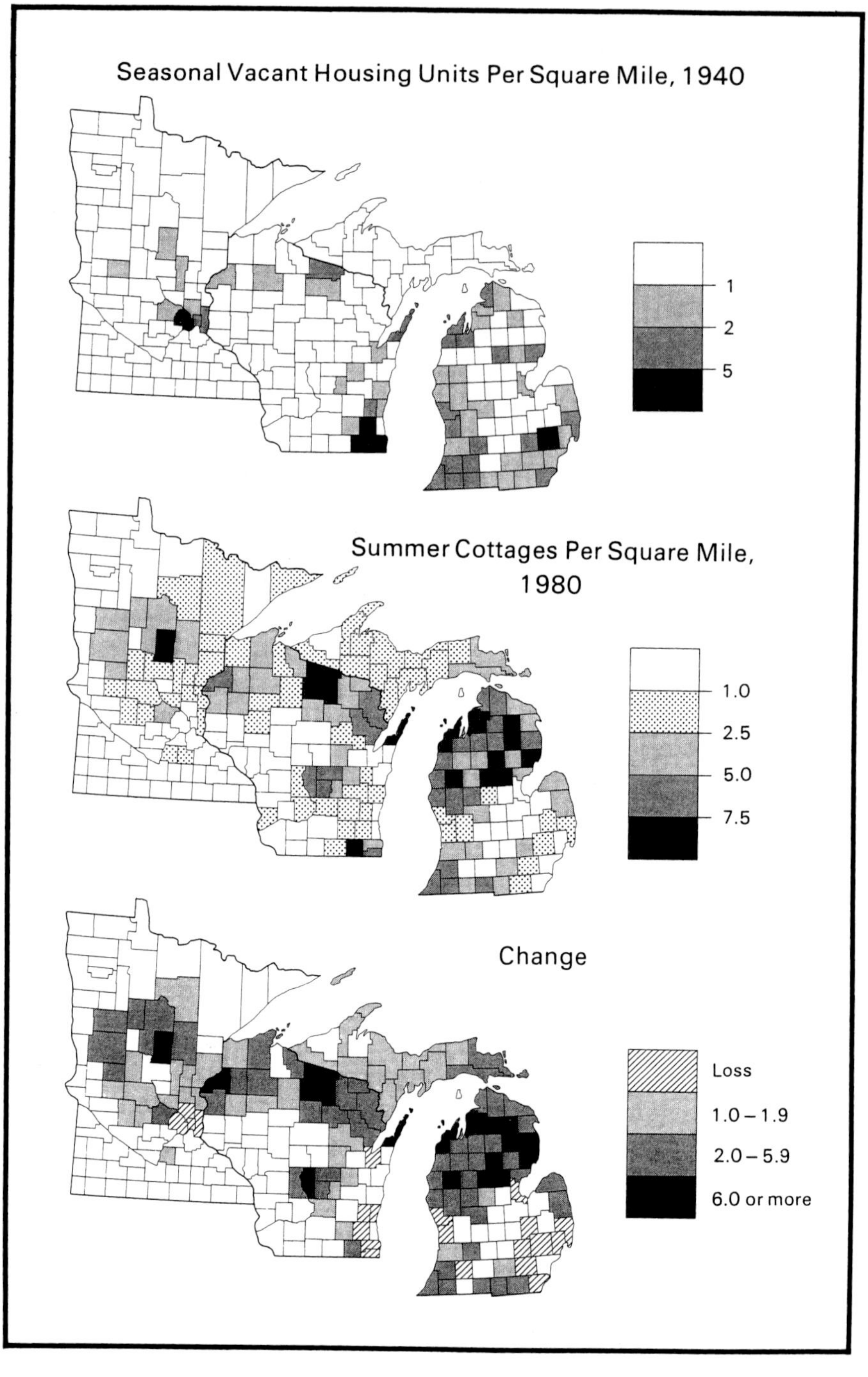

FIG. 10.2. Summer cottages in the Upper Great Lake states between1940 and 1980.

use." Housing units held for occasional use include apartments kept in city centers by well-to-do people whose primary residences are elsewhere. I corrected for these pieds-à-terre by subtracting one percent of the total number of housing units in each county in 1980 from the number of units held for occasional use, and added the remainder to the number of vacant seasonal and migratory units to estimate the number of summer cottages in the county.

In 1980, most counties in the northern parts of the Upper Great Lake states were included in a summer-cottage belt that thinned out toward the west. Most of the counties in the northern part of the Lower Peninsula of Michigan had more than five summer cottages per square mile, but only Crow Wing County attained this level in Minnesota. The leading counties in Wisconsin were Vilas and Oneida in the center north, Door in the northeast, and Walworth and Kenosha in the southeast. Close behind were Burnett in the northwest, Oconto and Marinette in the northeast, and Adams, Waushara, and Marquette in the center.

The data for 1940 and 1980 are not strictly comparable, but they do suggest that the number of summer cottages increased substantially in the more remote rural areas of the north while it was declining sharply on the metropolitan peripheries, where many erstwhile summer cottages apparently have been converted into year-round retirement homes or even into residences for commuters. Fifty miles simply did not seem as far in 1980 as it seemed in 1940! In outstate Wisconsin the largest increases were in the predictable pentad of counties—Burnett, Vilas, Oneida, Door, and Adams.

Persons per Housing Unit

A second approach to the identification of resort areas uses data on the number of persons per housing unit in each township (Figs. 10.3 and 10.4). In 1980, 308 (24.3 percent) of Wisconsin's 1,269 townships had fewer than two persons per housing unit, and ninety-eight (7.7 percent) had fewer than one. If a township has fewer than one person per housing unit, one might assume that some of these units were summer cottages that were not occupied on the first day of April, the official census date.

As late as 1970, an examination of the number of persons per housing unit probably would have been unpersuasive, because families were so large that their size more than compensated for the unoccupied units, and the number of

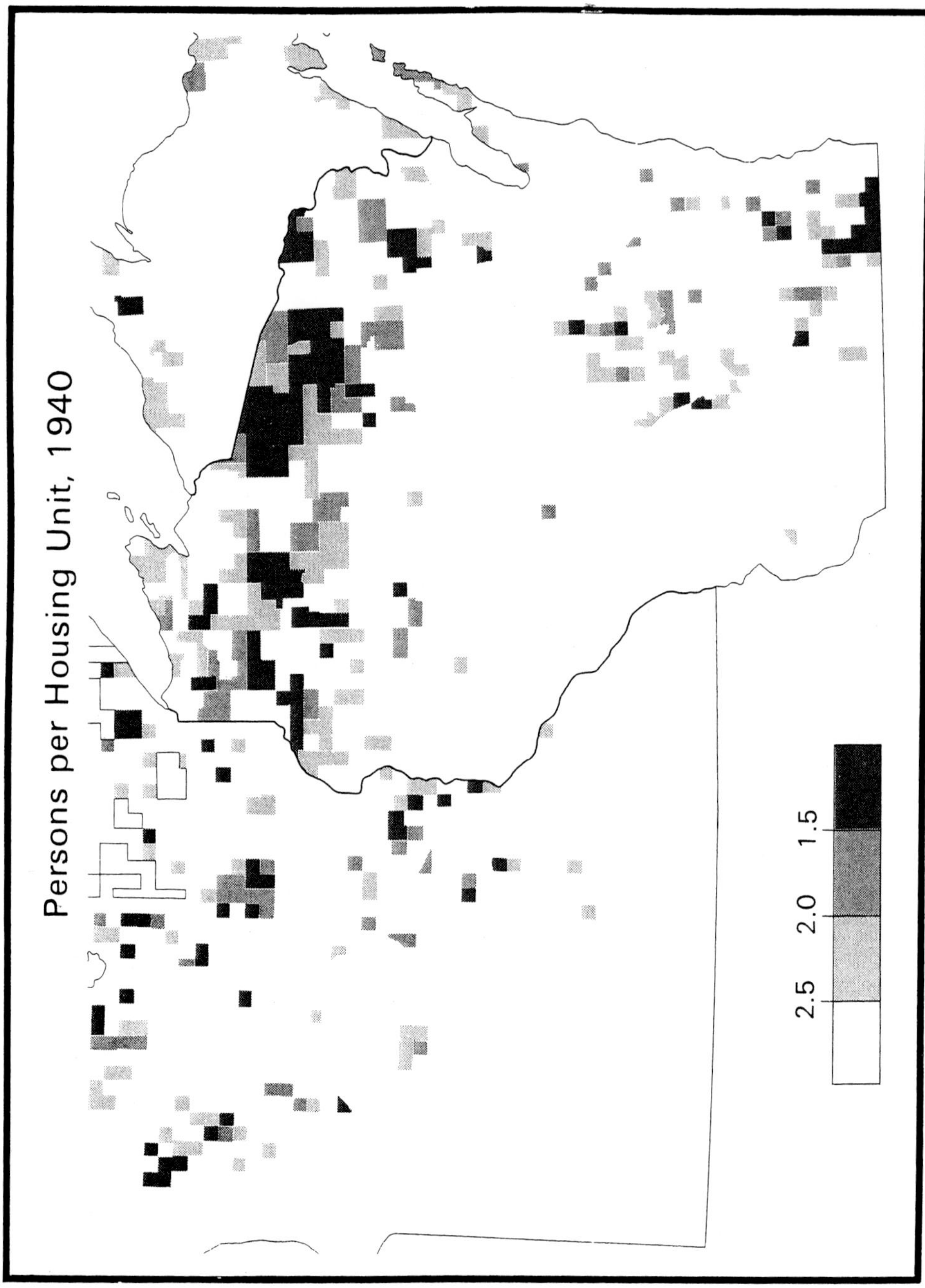

FIG. 10.3. Persons per housing unit, 1940.

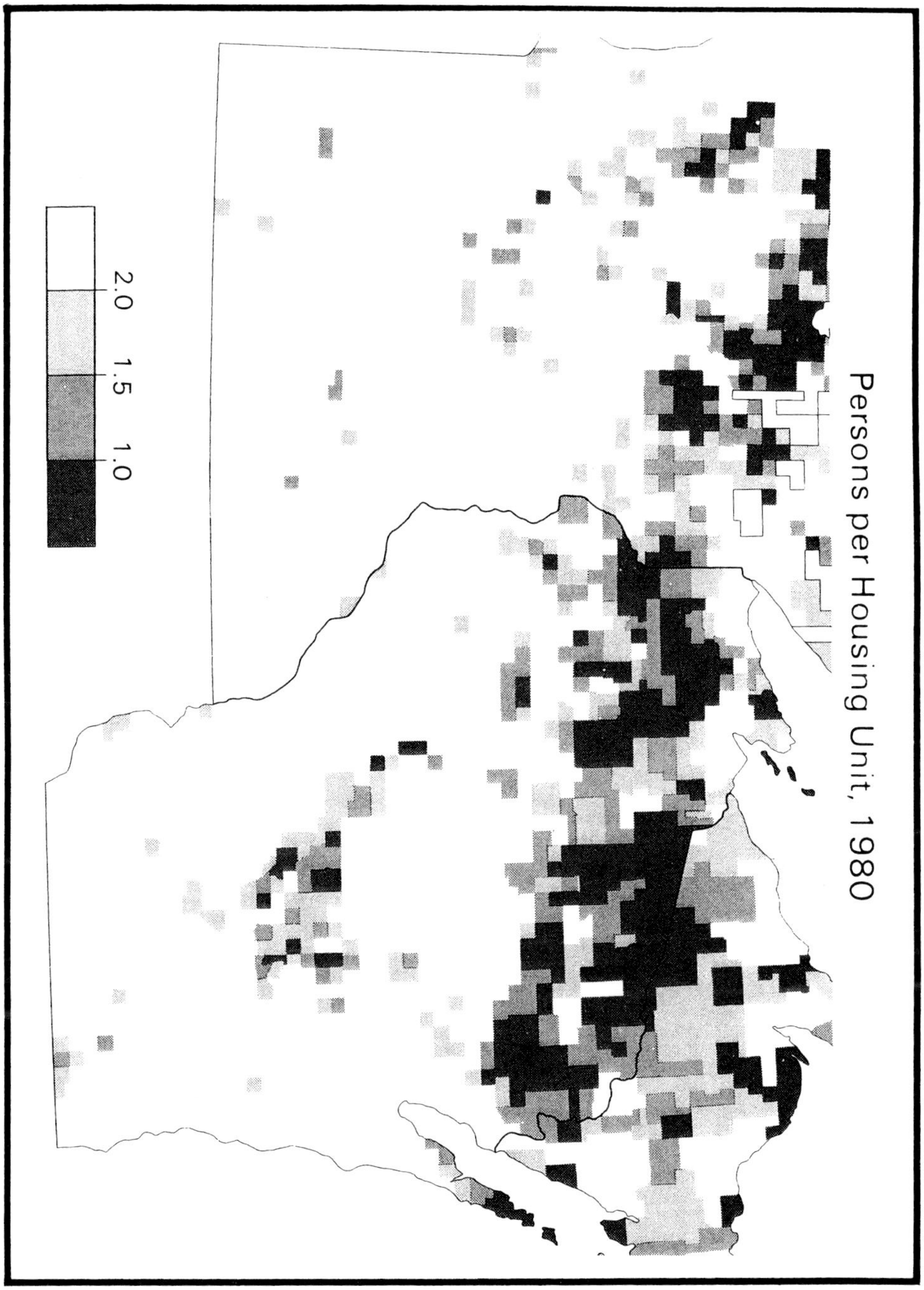

FIG. 10.4. Person per housing unit, 1980.

persons per housing unit in most townships was well above one. In the decade between 1970 and 1980, however, the number of persons per housing unit in Wisconsin dropped from 3.00 to only 2.52, nearly as much as it had dropped in the three decades between 1940 (when it was 3.50) and 1970, and the number of persons per housing unit became a useful indicator of summer resort townships that were sparsely occupied on the census date in April. This index is far less revealing at the county level, because only Vilas County in Wisconsin, no counties in Minnesota, and five counties in Michigan had fewer than one person per housing unit in 1980.

Most of the townships with fewer than two persons per housing unit in 1980 were in the lake-speckled morainic areas of the boreal forest (Fig. 10.4). Apart from the cluster on the sand plain of central Wisconsin, they were concentrated in a broad belt that stretched across the northern quarter of Wisconsin, tapered westward toward Mahnomen, Minnesota, and thence reached southeastward between Detroit Lakes and Alexandria.

The distribution of townships with fewer than two persons per housing unit in 1980 is so nearly similar to the distribution of summer cottages that fewness of persons per housing unit may be taken as an acceptable surrogate measure of summer cottages, and it provides far richer detail, because data on summer cottages are not available for areas smaller than counties. The number of persons per housing unit and the number of summer cottages per square mile in Wisconsin counties in 1980 had a respectable coefficient of correlation of -0.70, and this value increased to -0.76 if one excluded Walworth County, the essentially suburban area around Lake Geneva.

The townships with few persons per housing unit in 1980 were oriented toward bodies of inland water, but it would be quite unrealistic to expect the reverse, that all water bodies would be lined with summer cottages, because water bodies vary enormously in attractiveness and accessibility. Nevertheless, three-fifths of the Wisconsin townships with five or more percent of their area under inland water had fewer than two persons per housing unit in 1980 and fewer than two and a half persons per housing unit in 1940, and more than one-quarter of these townships had fewer than one person per housing unit in 1980 and fewer than one and a half persons per housing unit in 1940.

The index values for 1940 are one-half person per housing unit higher than in 1980 because families were larger, but, given this difference, the orientation toward bodies of inland water was even more pronounced. The boreal forest belt of townships with few persons per housing unit was still inchoate, although Wisconsin had harbingers in Burnett, in Vilas and Oneida, and in Door counties. There was a striking concentration southwest of Milwaukee near Lake Geneva. There were also scatters around Milwaukee and the Twin Cities that had completely disappeared by 1980.

Changes in the distribution of townships with few persons per housing unit reinforce the impression that many summer cottages, especially those within easy commuting distance of major metropolitan centers, have been converted into year-round residences, whether for retired people or for commuters. In the north, the number of townships with few persons per housing unit has continued to increase. In Wisconsin, cottage areas apparently have spread from their lakeshore nuclei into adjacent townships with little or no inland water.

Hotels, Motels, and Resorts

A third approach to the identification of resort areas uses data on the total number and gross sales of all hotels, motels, and resorts in each county in Wisconsin. These data have been extracted from the sales tax records of the state Department of Revenue by the Recreation Resources Center of the University Extension Service, which publishes an annual report on the numbers and sales of thirteen categories of retail and service establishments that are sensitive to purchases by nonresidents. In 1982, Wisconsin had 3,725 hotels, motels, and resorts, which reported gross sales of $486,794,525.

The raw data must be standardized to compensate for considerable variations in the areas of counties, the sizes of their populations, and the total volume of sales by all establishments subject to sales tax. The number of hotels, motels, and resorts in a county is partly a function of the size of its population because every place of any consequence has downtown hotels and suburban motels, and major cities, such as Milwaukee, Madison, Green Bay, Appleton, Oshkosh, and La Crosse, have conspicuous concentrations (Fig. 10.5). The number of hotels, motels, and resorts per square mile has a strikingly bimodal geographical distribution, however, and there are also massive concentrations in resort counties, such as Sawyer in the northwest, Vilas and Oneida in the center north, Door in the northeast, and Walworth in the southeast.

The counties with large numbers of hotels, motels, and resorts per capita are concentrated in the sparsely populated summer cottage belt across the northern

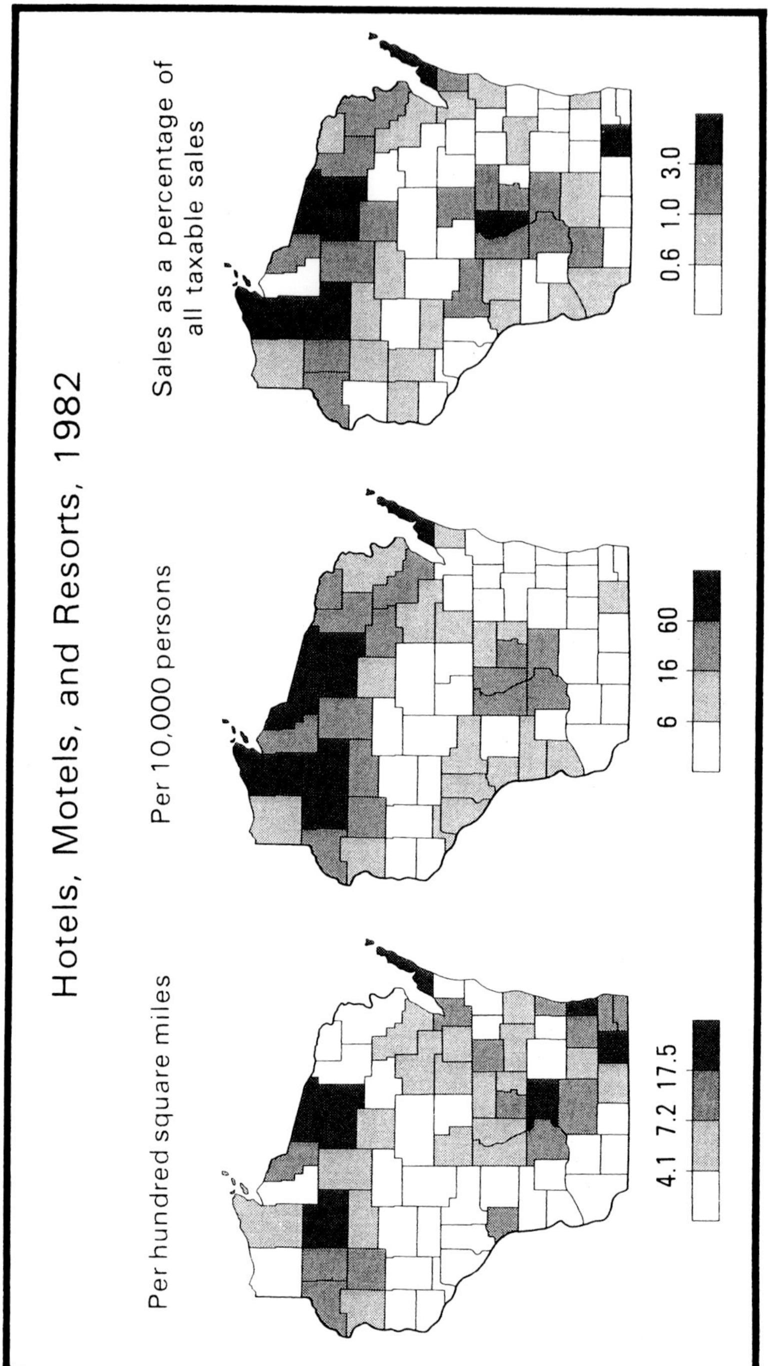

FIG. 10.5. Hotels, motels, and resorts in Wisconsin counties, 1982.

part of Wisconsin, with especially high values in Sawyer, Vilas and Oneida, and Door counties. The Dells, sand plain, and morainic areas in the central part of the state have a lesser concentration, and a string of counties on the western edge along the Mississippi River are above the state average. Hotels, motels, and resorts account for a significant percentage of all sales in many counties in northern and central Wisconsin. These data reflect not only the importance of tourism in these areas and the dependence of the local economies upon it, but also the lack of alternative economic opportunity and other sources of income in the sparsely populated boreal forest and the thirsty soils of the Central Sand Plain.

Differences Between Resort Areas

Three lines of evidence—the number of summer cottages, the number of persons per housing unit, and the number of and sales by hotels, motels, and resorts—identify five major clusters of resort counties in Wisconsin: Sawyer and Burnett with Bayfield and Washburn in the northwest; Vilas and Oneida with Iron in the center north; Door in the northeast; Adams, Marquette, Columbia, and Sauk in the center; and Walworth in the southeast. Resort areas are doubly different. They differ both from nonresort areas and from one another. Each resort area has its own distinctive character, and it appeals to distinctive personality types. For example, some resort areas appeal to white-collar types, who prefer outdoor activities that combine uplift and mind improvement with peace and quiet: nature study, photography, hiking, sailing, canoeing, cross-country skiing. Some resort areas appeal to blue-collar types, who want speed and blood: roaring powerboats, snarling snowmobiles, the excitement of hunting and fishing. White-collar types watch birds; blue-collar types shoot them. Some resort areas appeal to entire families with rides, shows, and other forms of commercial entertainment. Most resort areas have some mix of all three types of activities, but one clearly is dominant and gives the area its image.

I selected Door County, Vilas County, and Adams County for intensive investigation because they best illustrate the distinctiveness of individual resort areas. I used the yellow pages of the local telephone directories to identify the numbers of the different types of business establishments that cater to the tourist trade.

Door County

The Door Peninsula extends seventy miles northeastward from the city of Green Bay to the treacherous waters of the Porte des Morts ("Door of the Dead"), from which it takes its name. Door County occupies the northern two-thirds of the peninsula, but the local people know that "Door County" starts north of Sturgeon Bay, the county seat. The peninsula is a cuesta underlain by the resistant Niagara dolomite. The western side is a bold escarpment up to 200 feet high that commands splendid vistas across the water of Green Bay. The long lines of sheer, wave-cut cliffs at the base of the escarpment are interrupted by shallow pocket harbors. These harbors, sheltered by picturesque rocky headlands, are ideal for pleasure crafts and small fishing boats. The eastern side of the peninsula, which slopes gently beneath Lake Michigan, boasts belts of dunes and wide expanses of sandy beach.

The peninsula was settled by Scandinavians, who brought a tradition of log construction, including the use of "stove-wood" walls. There are many fine, old log buildings. The early settlers scratched out a living by farming the thin soil, but they had to supplement their income by fishing and lumbering. They cut cordwood for lake steamers and developed a brisk trade in cedar shingles with Chicago.

Regular steamship service to and from Chicago lured the first summer visitors, especially sufferers from hay fever, who boarded in private homes. By World War I, enterprising local people had started to supplement their income by establishing summer hotels. A railroad line was completed as far as Sturgeon Bay in 1894, but the thirty-five mile trip from Green Bay took three and a half hours and required a change of trains. The boat trip from Chicago took three days, with stops at Milwaukee and smaller ports along the way, and only people with long vacations could afford the time.

Many early visitors were affluent professionals or academics from Chicago or Milwaukee. These people sought sanctuary, not excitement, and they placed their stamp on the peninsula as a haven of peace and quiet where nature could be enjoyed and culture could be appreciated. Door County boasts the oldest summer-stock theater company in the United States, which dates back to 1935. The picturesque scenery and quiet lifestyle have attracted artists, who have established studios and galleries in quaint old log buildings. Door County has far more than its share of art galleries, gift shops, antique stores, and cutely named boutiques. It also has far more than its share of quiet recreational activities, such as golf, tennis, bicycling, and sailboating.

Snowmobiling was the first important attraction when winter tourism began to develop in the early 1970s, but it was not really appropriate to the area. By 1984, snowmobiling had dropped to a mere tenth of its peak popularity. It has been replaced by cross-country skiing, a quiet and peaceful form of recreation that is enjoyed by as many as 10,000 people on a good winter weekend. Door County claims to be one of the top ten cross-country ski areas in the United Sates.

Northern Door County has almost one eating place for every hundred permanent residents. A resort area is fortunate if it can cater to the gustatory needs of visitors with a local food specialty—or even better, a food ritual, such as a clambake, a barbecue, or a luau—that can be widely publicized as symbolic of the area. The specialty of Door County is the fish boil, an "ancient Scandinavian tradition" that seems to have been invented in the county early in the twentieth century. The impresario boils onions, potatoes, and whitefish in a huge kettle over an open wood fire. At just the right moment, he throws kerosene on the fire to make the water boil over, which removes the fish oil and edifies onlookers with a spectacular billow of smoke and steam.

Differences between Door County and Vilas County in drinking and lodging places reflect differences between white-collar and blue-collar resort areas. One would expect beer to be more popular in a blue-collar area, and Vilas County has more taverns than Door County, but about the same number of cocktail lounges, liquor stores, and nightclubs. Door County has more motels, hotels, and campgrounds than Vilas County, but Vilas County has more resorts. Door County also has far fewer group camps than Vilas County.

Vilas County

Few other parts of the world have as many lakes per square mile as the moraines and pitted outwash plains of Vilas and Oneida counties. Vilas County alone claims 632 "fishing lakes" and Oneida County has another 521. Twenty-eight of these lakes are linked by wandering streams to form the Eagle Chain of Lakes, which runs southward more than fifteen miles from Eagle River, county seat of Vilas County, past the village of Three Lakes.

Vilas and Oneida counties had splendid stands of white pine that attracted two waves of lumbermen. The first group began log drives on the rivers in 1856, but areas away from the rivers escaped until the early 1880s, when railroads were pushed northward into the area from the principal center at Rhinelander. The lumber companies built camps along the railroads for their workers, and the larger camps had primitive hotels for visiting company officials. Passenger trains from Chicago and Milwaukee made these hotels accessible to venturesome vacationers and sufferers from hay fever and asthma.

The logging boom was brief; it had almost ended by 1900, and the logging camps were becoming ghost towns. Some summer visitors bought the old company houses and moved them to lakeshore sites; others built their own rustic cottages. Local people first became involved in the tourist business by renting rooms to summer visitors. Then they expanded their operations by building separate cabins where visitors could sleep, but they still served all meals in the main house. This arrangement evolved into the idea of the "resort," with a main lodge serving scattered cabins. Everyone ate in the lodge, which had a large hall for dining, games, and social activities. The cabins were primitive. They had beds, but no running water or inside plumbing. Only the better cabins had kerosene lamps. Housekeeping cottages for summer rental were rare before World War II. The majority of the tourist accommodations in Vilas County still are called resorts rather than hotels or motels.

Hunting and fishing were the principal attractions of Vilas County in the early days, and they remain important. The easy violence of hunting and fishing was associated with a tradition of disdain for the legal niceties. The lumber companies frowned on drinking and prostitution, but many elected officials in Vilas County did not allow themselves to become unduly concerned with the strict enforcement of unpopular laws. Saloons operated openly during Prohibition, and nearly every tavern and resort had slot machines.

Affluent visitors began driving to Vilas County as early as the 1920s, but the automobile did not really supplant the passenger train until after World War II. Better cars and better roads brought the area within weekend driving time of Milwaukee and Chicago, and longer vacations encouraged regular visitors to consider building their own cottages instead of staying at resorts. Condominiums and family campgrounds also have become increasingly common, but the area within fifty miles of Eagle River still has more than 700 resorts, motels, and hotels.

The highways of Vilas County are clogged in summer with boat trailers, camper vans, and pickup trucks, all of which are uncommon in Door County, and the highways in winter are full of snowmobile trailers. Vilas County has far more businesses that deal with engines—powerboats in summer and snowmobiles in winter—than either of the other two resort areas. The tourist season in Vilas County used to run from the opening day of the fishing season in May until the end of the deer season in November, but the snowmobile has played a major role in converting tourism into a year-round business.

Oscar Eliason of Vilas County built the first snowmobile in 1924. The county has taken his invention to its bosom, because it has turned snow into white gold. The area is crisscrossed with clearly marked and carefully groomed snowmobile trails, of which St. German township alone has more than 500 miles. The first World's Championship Snowmobile Derby was an informal affair on the frozen surface of Dollar Lake in 1964, but by the 1980s 40,000 people show up in Eagle River on the third weekend of January for three days of snowmobile racing in the special stadium that has been built.

The center of Eagle River is like the centers of many other towns in resort areas, although on a somewhat grander scale. Many stores have been built in the "northwoods rustic" style of unpainted wood and cedar shakes. The kinds of business establishments one would expect to find on an ordinary Main Street have been shouldered off onto back and side streets, and both sides of the two principal blocks on the main street are given over to fairly high order specialty stores.

Adams County

Adams County is not one, but three quite distinct resort areas—the Dells, the sand plain, and the new lakes—that have little in common except the accident of geographic proximity.

At the southwestern corner of the county the Wisconsin River has carved the narrow, steep-sided gorge known as the Dells into a cross-bedded sandstone that forms sheer cliffs eighty to 100 feet high and stands in striking, isolated rock pillars. The Dells have been a mecca for tourists since 1848, when the geologist J. G. Norwood waxed rhapsodic about their picturesque "architraves, sculptured cornices, moulded capitals, scrolls, and fluted columns." The Dells have easy access from major metropolitan areas. Driving time is three hours from Milwaukee, and four and a half from Chicago or the Twin Cities, easily close enough for a family weekend trip.

Few other places in the United States have such intense concentrations of family-oriented commercial entertainment. A boat trip to see the gorge presumably is still a reason for visiting the Dells, but the natural scenic attractions of the area have been increasingly overshadowed by the garish signs and outlandish architecture of manmade marvels. The Dells is the kind of place that aesthetes deplore and people adore. It attracts two and a half million visitors each year, which is equivalent to slightly more than half the total population of the state of Wisconsin.

The Dells area is highly concentrated, and it has had almost no effect on the central sand plain of Adams County to the north. The sandy soils are excessively droughty, and they carry poor stands of jack and white pine. As late as 1960, much of the land could be bought for as little as $25 an acre, or even for back taxes. The Nekoosa Paper Company acquired 36,000 acres of land in the northern part of the county and is managing the land to produce pulpwood and sawtimber on a 45-year cycle. The paper company created the Wisconsin River Power Company and two "flowages" (reservoirs) for power production. The shores of these reservoirs have little recreational development and the power company still holds 8,000 acres of lakeshore land that it has never developed.

Four men from Chicago started the first flurry of resort activity in Adams County in 1926, when they platted the Dellwood subdivision on the Wisconsin River. They laid out streets and riverfront parks; built a hotel, a community house, and a pavilion; ran special weekend excursion trains from Chicago; and sold 4,700 small fifty-by-125 lots for $50 apiece. Most of the lots at Dellwood had been sold by 1929, but the Depression halted development and many lots had been sold for back taxes before better roads and automobiles made the area more accessible after World War II. Today Dellwood has approximately 400 simple cottages and 100 mobile homes. They have only limited legal access to the lake, and most are occupied for only two to four weeks in summer, plus an occasional long weekend.

Dellwood seems to have set the tone for resort development on the sand plain in Adams County, which has little to offer except cheap land under the pines. Many of the summer homes in Adams County are small owner-built cabins without running water or electricity, although mobile homes have become increasingly common. A high percentage of the mobile homes have been immobilized, and the owners have attached new rooms, sundecks, and other improvements.

The importance of Adams County as a resort area is not reflected in its business patterns. Although the county has more resort business establishments than nonresort areas, its numbers do not begin to compare with Door and Vilas except in family campgrounds, trailer parks, and RV parks. The sand plain is a haven for people who cannot afford more expensive resort areas, and they do not spend much money in the local area.

An exception is Rome, the northernmost township in Adams County, where three new water bodies were impounded—Lake Sherwood in 1968, Lake Camelot in 1970, and Lake Arrowhead in 1974—by a developer who had similar projects elsewhere. The land around the new lakes has been subdivided into roughly 2,000 half-acre lots, or a grand total of 6,000 new lakeshore lots.

Lake Arrowhead has a handsome, three-story clubhouse that would be the envy of many a country club. Golf courses and tennis courts are being added. The large thirty-five-by-seventy–foot swimming pool is next to the clubhouse, and a smaller pool is at the ski chalet, which has a rope tow and short runs. The lake has been stocked with fish; one beach house and docking facility is already in use, and a second is under construction. Lot owners at Lake Arrowhead are mainly from Milwaukee, southeastern Wisconsin, Chicago, or northeastern Illinois. Their average age is thirty-five to fifty, and their annual income is in the range of $22,000 to $30,000. These new summer people are several economic notches, at least, above those who have been summering in the county for a longer time. A few businesses have begun to cater to their needs, but this development is still quite embryonic when compared to other resort areas.

Epilogue

The basic conclusion of this study is quite simple: resort areas are not all alike. They differ greatly from nonresort areas, and they also differ from each other. A resort area cannot be all things to all people, and there really is no reason why it should try. A visitor to the "wrong" type of resort area could feel distinctly uncomfortable. Many bird watchers, nature lovers, canoeists, and cross-country skiers would be offended by the atmosphere of Vilas County. Conversely, hunters and snowmobilers are tolerated but not especially welcome in Door County, and well-to-do people would be scornful of the camper trailers that dot the landscape in Adams County.

A final observation is engendered by the phenomenal development in the northern part of Adams County, which would have been impossible without a massive injection of outside capital. Some rural areas in the United States seem to change only slightly over the decades, whereas others seem to change almost overnight. Much of the difference appears to result from massive external investment. Students of rural America need to identify and focus on the areas of rapid change, and they need to know far more about what attracts the attention and interest of large-scale investors.

Mobile Home

CHAPTER 11

Parks*

Virtually every place in the United States, no matter how small, has a cluster of mobile homes out at the edge of town. Larger places have more and larger clusters, many formalized as mobile home parks. Clusters of twenty or fewer units are generally fairly informal, with unpaved streets, few facilities, and haphazardly sited units. Parks with fifty or more units are generally planned and laid out more meticulously, with more order, more structure, more sense of a strong controlling hand. Significant economies of scale start to kick in somewhere between thirty and 100 units, which is the minimum size to warrant a full-time resident manager.

Mobile home parks vary enormously, from simple utilitarian parks with long, harsh rows of gleaming metal boxes lined up with military precision to carefully landscaped upscale mobile home communities that would put most suburbs to shame. Their range of quality has widened enormously in recent years. The best have become very good indeed, while the poorest remain ominous and repugnant, and a few are so threatening that we feared to venture into them.

Most of the newest and best upscale mobile home communities are in well-to-do retirement areas in Florida, California, and Arizona. Utilitarian parks are everywhere, wherever low-income people need inexpensive (and quick) housing: on the fringes of cities and towns, in rural areas for migrant farmworkers, near construction sites and industrial plants, especially meat

---

* Part Four, "Parks," in John Fraser Hart, Michelle J. Rhodes, and John T. Morgan, *The Unknown World of the Mobile Home* (Baltimore: The Johns Hopkins University Press, in association with the Center for American Places, 2002).

packing plants, near military bases, and as emergency housing where disasters have decimated the housing stock. They have not strayed far from the old, original, easy-in easy-out layout, with long straight lines of long rectangular boxes.

Mobile home parks are unusual residential areas in that the residents own their homes but do not own the land on which their homes are sited. They pay the park owner a monthly fee to cover lot rental, services, and facilities. They pay personal property taxes on their homes, but the park owner pays the real estate tax on the land. Most smaller parks are independently owned and managed, but larger parks may be owned by mobile home dealers, by manufacturers, by national chains, or by institutional investors.

In the mid-1990s, the managers of real estate investment trusts (REITs) decided that mobile home parks with 100 spaces or more were a better investment opportunity than apartment houses. Management and maintenance expenses were modest, rents were a regular and reliable source of income, and mobile home parks had low turnover rates because of the high cost of relocating units. It is hard for delinquent tenants to skip out from a mobile home park without paying their rent, as they can from an apartment, and the management can slap a lien on the units of tenants who are too delinquent. Mobile home parks were an especially attractive investment near cities, where land prices were so high that individuals could not afford to buy lots on which to site their units, and in the year 2000 it was estimated that the ten largest mobile home REITs controlled two percent of all mobile home parks. In 1996, ROC Communities, one of the larger REITs, owned and managed 110 parks with 28,000 rental sites.

Critics have had a field day cataloging the problems that residents have had with the owners of older utilitarian parks, especially in urban fringe areas. Inflation and rising taxes will inevitably force even the best-intentioned park owners to increase their lot rents over time, and retired people on low fixed incomes will just as inevitably complain bitterly about any rent increase.

Legally, mobile home parks are the private property of their owners, who are free to impose whatever rules and regulations they wish, and to change them whenever they see fit to do so. In the old days, parks for transient trailers had strict rules to protect the other residents as well as the park owners, and that tradition persists in some contemporary parks, especially where park vacancies are scarce, as they are in many urban areas.

Utilitarian mobile home parks perforce are semipublic places. Lots are so small that parks have less privacy and far more social interaction than conventional residential suburbs. People are friendly and curious, and most parks seem to be genuinely caring communities. Life is in the narrow streets, which are lined with parked cars, many with hoods raised, seeming to devour the young men half-buried in their engines. Many of the residents are skilled artisans who enjoy doing things with their hands, such as maintaining and adding on to their units, and they are willing to help out on repair and maintenance jobs in the park. "Anything that goes wrong in the park can be fixed by somebody who lives here," we were told more than once.

Upscale mobile home communities are a relatively recent development. They are designed to be more attractive than utilitarian parks, most of which have become permanent residential areas but still retain the traditional cramped layout of parks for transient travel trailers, with small lots and few amenities. The people who buy expensive new mobile homes as permanent residences want nicer places to put them, and the new upscale parks are laid out like conventional residential subdivisions, from which they are distinguishable only by the nature of their housing stock. In turn, the availability of nice places to put them has encouraged more people to buy newer and more expensive models of mobile homes.

Handsome new mobile homes are especially attractive to families who wish to move from rental apartments, to empty-nesters who wish to sell their site-built houses and downsize, and to retired people, many of whom want affordable second homes in the Sunbelt. Most of the new upscale mobile home communities are in Sunbelt areas, although some have also been developed near major cities in other parts of the country.

The developers of the new upscale mobile home communities are targeting well-to-do consumers with an impressive range of amenities. Many focus on golf courses and marinas. Their clubhouses may have restaurants and sandwich shops, post offices and convenience stores, card and billiard rooms, dance floors, and saunas and jacuzzis in the exercise rooms. Next to the clubhouse may be the heated swimming pool, tennis and shuffleboard courts, putting greens and bowling greens, croquet courts, a horseshoe pitch, and perhaps even a softball diamond.

Many of the new parks have guarded gatehouses with twenty-four–hour security, which senior citizens especially appreciate. The curving streets are paved and curbed, with distinctive lights and signs, and the park maintains the lawns meticulously. The lots are large enough that each home has a two-car parking apron and a carport on one side, and a patio with privacy fencing on the other.

Many of the residents of the new upscale parks in the Sunbelt are "snowbirds" who simply lock up the place in the spring and drive to their summer homes up north. They came initially to escape the northern winter, but in the park they have developed friendships that keep drawing them back. They have large amounts of discretionary time, but the park keeps them fully occupied with a frenzied social whirl of club meetings, dances, classes, tournaments, potluck meals, and similar group activities.

Individual Parks

Each utilitarian mobile home park and each upscale mobile home community has its own distinctive history, and its own demographic and socioeconomic profile. We have sampled a number of individual parks to give something of the flavor of their uniqueness and diversity.

Southwestern Kansas

At the eastern edge of Garden City, Kansas, a strip of low, bushy austrees north of US 50 tries to screen one of the largest mobile home parks in the United States. This park has 600 lots, and the owner plans to expand to 668. The lots, each measuring fifty by 100 feet, are large enough to hand double-wide mobile homes, but only a handful of the units are double-wides. Only ten lots are empty, and the park has a waiting list, even though it is grimly utilitarian, with units rigidly aligned in long, monotonous, parallel rows of thirty-seven. This park is merely the largest of many that serve the newly emergent meat-packing complex of southwestern Kansas.

Since 1960, southwestern Kansas has become one of the nation's leading beef-producing areas. In the 1880s, local farmers began to divert water from the Arkansas River to irrigate the level bottomlands along the stream, and they grew sugar beets as their principal cash crop. Farmers on the dry upland grew winter wheat, and they ran herds of beef cattle on the rangeland that was not suited to cultivation. In the 1950s, sugar beets were becoming less profitable, and some local entrepreneurs developed feedlots, where they could bring lean cattle from the range and fatten them to market weight.

A few farmers pumped from shallow wells, but they irrigated most fields only by gravity until the 1960s, when the development of powerful turbine pumps enabled them to pull water from the famous Ogallala Aquifer, 200 feet below the surface. The development of center-pivot sprinkler irrigation systems enabled them to irrigate the rolling uplands, where they grew bumper crops of corn and alfalfa. The abundant supply of feed in the area encouraged the development of huge feedlots for beef cattle, and, by 1990, southwestern Kansas had nearly 100 feedlots with a total capacity of more than one million beef cattle.

The availability of fed cattle, plus an abundance of water from the Ogallala Aquifer, attracted meat-packing companies to the area. In 1980, IBP, Inc. (formerly Iowa Beef Processors), opened the world's largest beef-packing plant in Holcomb, ten miles west of Garden City. The Holcomb plant employs 2,800 workers, and in 1983 the Monfort packing company modernized its Garden City plant and doubled its work force to 1,300. These two plants slaughter 8,400 cattle a day. In addition, Dodge City has two large beef-packing plants, Liberal has one, and Guymon, just across the state line in Oklahoma, has a major hog-slaughtering facility.

The new plants in southwestern Kansas highlight how dramatically the meat-packing industry has changed since World War II. It has moved from metropolitan areas to smaller places that are closer to the source of the animals, and it has replaced union butchers with lower-wage, unskilled workers who constantly perform the same, simple, monotonous tasks on production lines. Carcasses or boxes roll inexorably past them at a rate of around 400 an hour, or one every nine seconds for the entire working day, and they must do precisely the same thing to each one.

Work in a meat-packing plant has always been brutish, poorly paid, and extremely hazardous. Meat packing is second only to underground mining as the nation's most dangerous industry. The knives and saws that slice up the animals are equally ready to butcher human flesh. The packing companies have been criticized so harshly for so long that they have become skittish, and they are understandably reluctant to answer any questions from outsiders. One company executive even refused to tell us the number of cars that could be parked in his company parking lot.

The meat-packing companies had to recruit new workers for their plants, because southwestern Kansas is sparsely populated and few local people were willing to take such dangerous jobs at such low wages. The companies recruited so well that the population of Garden City town and township boomed from 20,980 people in 1980 to 29,565 in 1990, with most of the growth in the first half of the decade, when Monfort doubled its work force. Some of the new people were Anglos, some were immigrants from Mexico, and some were refugees from Southeast Asia, mostly Vietnamese and Laotians. Minh Duong, personnel director of the IBP plant in Holcomb, said,

> "We started construction in 1979 and began processing in 1981, but this area had no housing for workers. Some of them lived in parking lots, and some camped out in parks. We had to delay starting a second shift until housing was available in a large trailer park east of Garden City. It is ten miles from our plant, but the land there was cheap. I have heard that they bought an old dump and leveled it."

IBP encouraged contractors to develop trailer parks by guaranteeing them referrals of our employees. At one time we deducted their rent payments directly from their paychecks, but that sort of died out around 1990 because we had enough housing. Most of our employees rent-to-own their units. The average price of a unit is around $14,000, and they rent a space in a park for $135 a month. Garden City has four trailer parks, and it doesn't want any more, so now IBP is trying to encourage contractors to build apartments that our employees can afford.

Rob Martin owns and operates the huge East Garden Village Mobile Home Park east of Garden City. "My parents have been in the 'wobbly box' business for many years," he said. They gradually built up a 520-unit park in Sioux City, Iowa, which is right across the Missouri River from IBP headquarters in Dakota City, Nebraska. When IBP came to Garden City, which needed housing badly, its managers asked Rob's father to co-venture with them. IBP put up eighty percent and promised to buy 500 units in increments of 125. Then the company was bought out, and IBP cut the program.

Rob built the park up to 280 units by using a variety of incentives. He paid the cost of moving a unit from anywhere within a fifty-mile radius. He offered cash incentives and allowed lots to be occupied rent-free for the first month. He gave a month's free rent to anyone who got a friend to move in, but he does not need to do that anymore. In 1986, he bought the 125 IBP units. He rented them until 1992, because he needed the cash flow. He had the park full by 1993 and plans to expand to 688 lots. He is going to develop a retail complex near the highway with a convenience store, a liquor store, and a car wash, and will have space to rent.

Rob is developing a second large mobile home park near Guymon, Oklahoma, where the Seaboard Corporation is developing a huge new hog production complex. The Country View Mobile Home Park southeast of Guymon has the austere functional beauty of a place that is precisely what it says it is: an oasis of affordable housing for low-income people. At first glance it looks pretty bleak, with long straight rows of identical Masonite boxes set on bare seed-speckled ground with not a tree or even a bush in sight. "I didn't have time for any landscaping," Rob said, "because Seaboard wanted 'em here in a hurry, and I had to do 'em quick. I am going to irrigate the whole park, and then I am going to put in sidewalks. I have tried to plant trees many times in the Garden City park, but the kids kill them as fast as I can plant them."

In Guymon, Rob is also developing Country Estates Mobile Home Park, with 100 lots for double-wide units. He said:

> "The workers can afford them but they can't afford the houses that some contractors are building. I am often asked if I am interested in developing new parks, and I'm still in the market if I can find a good opportunity. I need an outlying area close to city utilities, flat, and with room enough to expand. Seaboard wants me to put 100 units in Ulysses, Kansas, where they have a lot of hog farms. They want me to put in smaller parks in smaller places for their farmworkers."

Rapid industrial and population growth often create a need for quick and inexpensive housing. Rob Martin is the largest developer of mobile home parks in Garden City and Guymon, and he has done a good job, but he is not an anomaly. Massive mobile home settlement and park development have accompanied the rapid growth of industry in rural areas through the South and West.

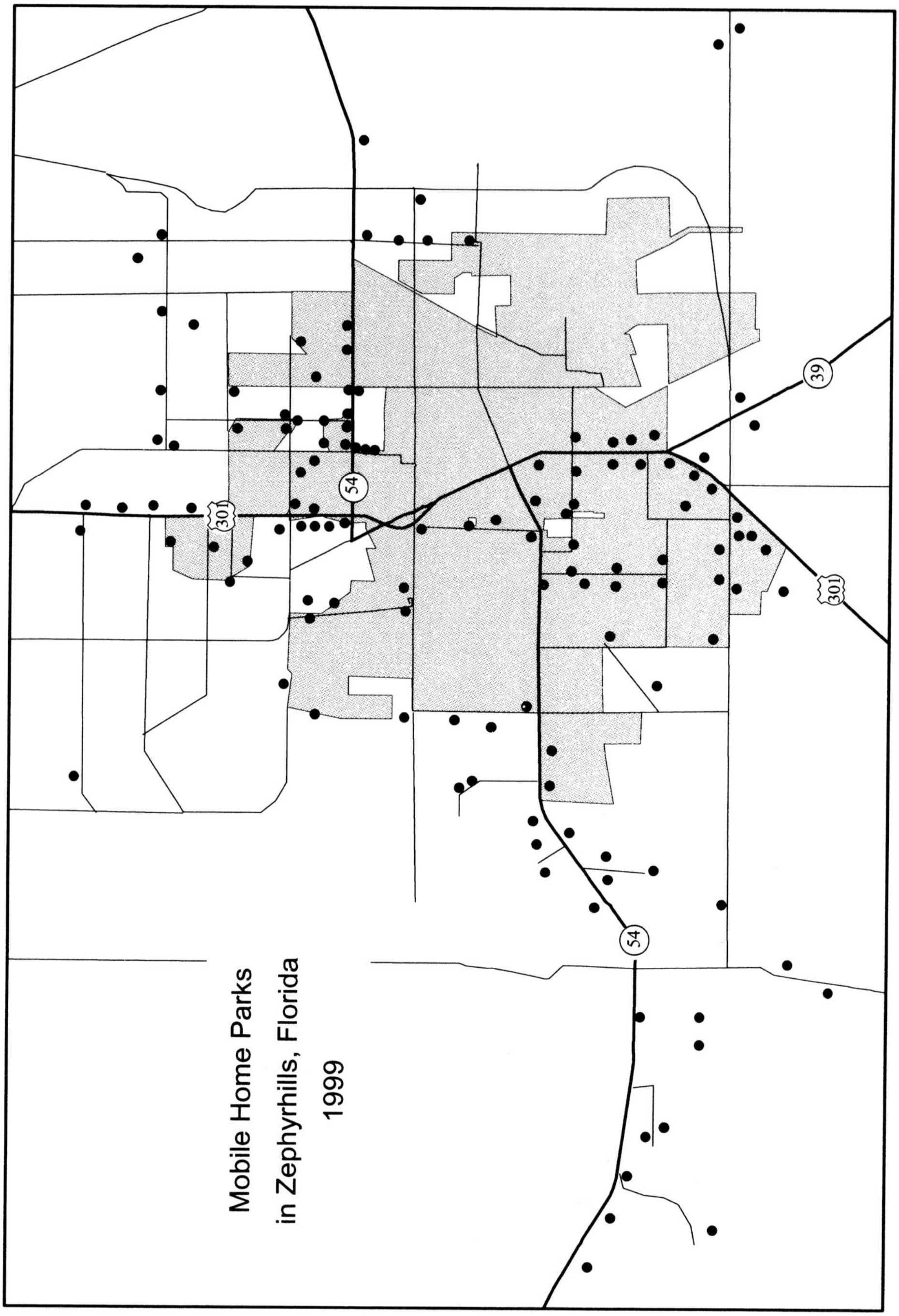

FIG. 11.1. Mobile home parks in Zephyrhills, Florida, 1999.

Florida

The mobile home is widely perceived as a perfectly acceptable form of housing in Florida, which has more mobile homes per capita and more per square mile than any other part of the United States. In 1990, the state had 821,000 mobile homes, which comprised 13.5 percent of its total housing stock, or more than one of every eight housing units. In rural Florida, more than one of every four housing units was a mobile home, and, in rural areas north of Orlando, the figure was closer to one of every two. Mobile homes in rural areas are generally single-sited, but in built-up areas they are clustered in parks.

The mobile homes and mobile home parks of Florida are concentrated on the west coast from north of Tampa southward to Naples. The Silver Coast, as this area is called, is quite unlike the Gold Coast on the east, which is livelier and better known. The Gold Coast attracts well-to-do people from the cities of the Northeast, people who are willing to live in high-rise apartment complexes. The Silver Coast, which is lower key and lesser known, takes its name from the color of the hair of most of its residents. It attracts retired people from the Midwest, the rural Northeast, and eastern Canada, people who prefer to live in single-family homes with yards. Every sixth dwelling unit on the Silver Coast is a mobile home, and most of the mobile homes are in parks of all sizes and price ranges, from the luxurious to the strictly utilitarian.

The Zephyrhills area, twenty-five miles northeast of Tampa, shows why trying to obtain accurate information on the numbers and locations of mobile home parks is like trying to pin down Jell-O. Zephyrhills is a winter refuge for people who have retired on modest incomes. The center of town, where Florida 54 takes leave of U.S. 301, is dominated by a large white store with huge signs in bold red letters advertising mobile home parts and accessories, such as patio covers, screen rooms, hurricane awnings, windows, doors, and aluminum siding and skirting.

On November 5, 1998, the "Welcome Back" edition of the *Zephyrhills News* published a map and directory listing 145 mobile home and RV parks, but giving no indication of their size. The Zephyrhills Chamber of Commerce said that seventeen parks had more than 250 spaces each, and the largest park had 994 spaces.

Some local enthusiasts boasted that the area actually had 170 parks, but the Zephyrhills (Fla.) 1:24,000 topographic quadrangle last updated by the U.S. Geological Survey in 1987, showed only 27 "trailer parks." The topographic map did not separately identify parks that were contiguous, as some are, but the discrepancy, even given the difference in dates, suggests that many of the smaller parks had only a handful of spaces (Fig. 11.1).

The 1990 Census of Housing reported that Zephyrhills had 6,000 mobile homes (50.1 percent of all housing units), and the adjacent rural areas had 6,750 more (66.7 percent of all housing units). The 1990 Census of Population found that Zephyrhills had 17,300 people, and the adjacent rural area had 18,300 more. Local enthusiasts claim that the population swells to 80,000 people in winter, but 65,000 to 70,000 is probably closer to the mark, unless some housing units are awfully crowded. There can be no doubt, however, that lots of snowbirds flock to Zephyrhills in the wintertime.

The snowbirds start moving in around October 15, but the big influx is after the Thanksgiving holiday. Some drive down early and then fly home for Thanksgiving. In early November, many church notice boards in Zephyrhills sport large "Welcome Back!" signs and business owners start blowing the dust off fixtures in restaurants and other establishments that they had closed for the summer. In spring, there's always a big discussion about whether to pull out before or after Easter, because people are always worried about getting caught in a late snowstorm while driving north.

One park in Zephyrhills had 290 lots that rented for $150 a month for the full year, and 240 lots that sold for $12,000 to $14,000 each. The park had two new units for sale at $60,000 each, but you could buy a used unit for anywhere from $3,000 to $30,000. The rental lots had septic, water, sewer, and garbage services, but lot owners had to pay for these services. Residents were allowed to have one visitor, who could stay for no more than thirty days, and they were charged $3 a night for each additional visitor.

About sixty residents lived in the park year-round, but most of the rest arrived around the first of November and headed off around the end of March. License plates are a poor indicator of their origins, because most preferred to buy cheap Florida plates, and some have elected to establish an official Florida residence because the state has no state income tax.

Two-fifths of the residents were from Michigan, one-fifth from New York, one-tenth from Canada, and the rest from all over, from states as far away as North Dakota and Maine. Michiganders were so numerous that they held an annual Michigan Picnic, to which other residents responded by organizing an annual Everybody But Michigan picnic two weeks later.

The park's residents seemed hyperorganized, with nearly 100 officers to manage a frenetic round of social activities. In addition to the full panoply of traditional officers, it had directors for bingo, bridge, canasta, shuffleboard, arts and crafts, blood pressure checks, soup day luncheons, potluck dinners, pancake suppers, greeters, coffee pourers, doughnut servers, song leaders, ticket sales, and hall decorations.

"We all know each other" was a common theme in the residential retirement parks on the Silver Coast, which are distinctly different from rental parks. Rental parks own their units and rent them to tenants. In residential parks, the residents own their units and rent or own the land on which their units are sited. The tenants in rental parks generally seem to have less interest in the maintenance and appearance of their units and parks, and these parks are more likely to be junk-strewn and unsightly.

Rental parks have an adversarial atmosphere, and tenants and managers alike seem to have chips on their shoulders. The tenants suspect the manager of rent-gouging, and the manager is afraid that tenants are going to skip out without paying the rent. In one rental park, we asked a woman where we could find the manager, and she snarled, "You don't want to live in this place." The managers are equally suspicious of inquisitive strangers, and we generally hit a blank wall when we tried to talk to them.

Residential parks are quite the opposite; if anything, they are almost too friendly. Most of their residents are retired people, and they have plenty of time on their hands. Most have been coming back to the same park each winter for years, and the residents of many retirement parks are also like extended families. The people who manage retirement parks must enjoy interacting with other people, and they are delighted to talk to inquisitive strangers, who just might be prospective residents.

Patricia Smith, fifty-nine, is the vivacious, gregarious, and tart-tongued resident manager and den mother of the K & K Mobile Home Park in Bradenton, an inexpensive park for older people who have retired on fixed incomes. The owners bought the park in 1976 as an investment property. They also own a park in Tavares, Florida. Pat came here in 1995, replacing a manager "who had been here since dirt was invented." The park has 175 mobiles. Fifty or so are occupied by year-round residents and the rest are seasonal, owned mainly by people from Michigan, New York, and Ohio.

"This is a great place for older folks and Canadians," Pat said. "It's an inexpensive way to live. Upkeep is easy and cheap compared to a house, and in a house they would be targets for every glib salesman that comes along." Old people want security, and the park is fairly safe. In houses they would be all alone, but in the park they look out for each other, and they get together to do jobs.

The residents are like-minded people, with a strong sense of community, and they have an active Park Tenants Association and clubhouse. Many residents are related, and in the summer up north some even get together for picnics. They have the same ailments, and they like to talk about their operations and prescriptions. "I have heard about so many knee and hip operations," said Pat, "that I think I could even perform one myself."

The park is a neat and tidy place. "It's their own hard-earned money," said Pat, "and they care." She has regular contests and awards prizes for the most attractive lots, and the Park Tenants Association helps her to enforce the park rules and regulations about maintenance and upkeep. Some residents are so old that they simply cannot maintain their lots, and the park has to do the work for them and charge them for it.

Screening new residents is a major responsibility and challenge, and Pat warns residents who sell their mobiles that the buyer they find might not be permitted to live in the park. At least one resident of each mobile must be fifty-five or over, and the park requests that all resident spouses and children be at least forty-five, "but there's really nothing that I can legally do about it, " said Pat, "if some old geezer decides to marry an 18-year-old bimbat."

Residents may have guests, but for no more than fifteen consecutive days or more than thirty days a year. Neither is considered a guest if two people are living together but not legally married, although both must be 55 or older. Some older couples elect to cohabit without benefit of clergy because they would lose their benefits if they remarried. The residents of the park are of a generation that frowns on such behavior, but Pat said that the statutes of the state of Florida prevent her from asking to see a marriage license.

In 1999, she was a bit uneasy about the immediate future, but she thought things would improve in a few years when the members of the baby-boom generation started to retire. The average age of her current residents is seventy-two, and "I keep a stack of sympathy cards on my desk," she said. "The kids don't want to live here when a resident dies. They clean out the place and sell if they can, or simply turn the title over to the park." Pat renovates the unit and tries to sell it, but rents it if she cannot sell it.

Pat is going to have to rent twelve to fourteen mobiles for the next few years, but she looks forward to the time when the baby-boomers start to arrive. "Nobody is building mobile parks any more," she said. Some of the baby-boomers are going to need inexpensive housing because they can't all afford luxury units.

The Six Lakes Country Club in North Fort Myers is an upscale, if not luxurious, mobile home park that laps around a private eighteen-hole golf course whose fairways are speckled with the six lakes that give the park its name. The paved streets that curve gently through the park are named for such illustrious golfers as Snead, Hogan, Nicklaus, and Palmer.

Johnny Johnson, a produce farmer from central Florida, developed the park in 1974. He was an avid golfer, but he knew where the money was, so he laid out a mobile home park around his golf course. He built a handsome office and recreational hall, an attractive sandwich stand and pro shop, a swimming pool, a putting green, and several shuffleboard and tennis courts. He laid out 595 lots, mostly sixty by eighty feet, of which 157 were on the fairways. Within a few years, his staff had sold all of them.

In 1984, the residents of the park formed a co-op and bought the park for six million dollars, which they now say was the smartest thing they ever did. Their numbers included ample legal and professional talent to handle all details of the transaction. The co-op offered shares at $10,000 apiece to raise the money to buy the park, and expected each resident to buy one share. By 1998, the value of a share had risen to $16,180, and only eight residents did not own one. The co-op does not like to own shares, but it did hold eight in 1998. The residents own their units, and they lease their lots from the co-op. In 1998, their assessments were $2,549 for lots on the golf course and $2,223 for other lots.

A high wall encloses the park, with a gatehouse at the principal entrance and narrow hole-in-the-wall side entrances. All gates are closed from six in the evening until six in the morning for security. The park has had three minor break-ins in the last five years, but nothing major, and it is a safe and comfortable place to live.

Six Lakes is an extraordinarily friendly place, as friendly as a Rotary Club meeting. Everyone smiles, everyone waves, everyone says hello. Everyone looks well fed and freshly scrubbed. The park manager is E. Bard Rupp, fifty-nine, genial, and avuncular. He taught school in Pennsylvania for twenty-five years before he moved to Florida. He is professionally licensed as a manager of community associations and is essentially the city manager for a "town" of around 1,200 people. He is responsible to a board of seven directors who are elected by the residents.

The park has a separate corporation for social affairs and its own community TV channel, which it uses for announcements. It runs the Sand Wedge Shop as a nonprofit operation to keep prices low because the residents want it that way. The park has eighteen full-time and twenty-one part-time employees, including a golf pro and a professional greenskeeper. It provides uniforms for its employees and has a good benefits program for them.

Residents must trim their shrubs, but the park mows all of the grass and maintains ditches and roads. It owns three water wells. Originally, it had a package plant to treat sewage and used the effluent water in a computerized irrigation system on the golf course, but in 1993 it hooked on to the North Fort Myers water and sewer service system. The park provides water and utilities, for which it charges residents $30 a month.

The average resident has been in the park for fifteen years or so. They come from all over, and from all income levels. Many could afford more expensive housing, but they prefer a small home that is easy to maintain. Forty percent of the residents are gone in the summer. Once they came in January, but now they come in October, and they do not leave until May.

Bard told me that the clubhouse building is officially certified as a hurricane shelter, and one time last year it housed 170 people. His pager went off as we were ending our conversation. The manager of the Sand Wedge Shop reported that she was out of beer, so Bard had to mount his golf cart and ride off to fetch a full keg for her.

Back Along

CHAPTER 12

the Milky Way*

Specialization has replaced diversification as the primary goal of good farm management in the latter half of the twentieth century. Once upon a time, American farmers tried to avoid "having all of their eggs in one basket" by doing a little bit of everything, but modern farmers, to an ever-increasing degree, have had to concentrate on doing what they do best—and doing it even better. They have spun off their less profitable and less efficient operations to other farmers and to other areas.

Dairy farming neatly illustrates the increasing specialization and geographic concentration of American agriculture. In 1924, almost every farm in the northeastern quadrant of the United States had a family cow or two, but by 1982 it was cheaper for many farmers to buy their milk at the supermarket than to bother with a cow, and dairy cows were concentrated in a few favored areas (Figure 12.1). The milky way that once extended from Minnesota to Maine had coagulated into scattered clots in response to different environmental constraints and economic opportunities. These differences are highlighted by dairy farms in Calumet County, Wisconsin; Madison County, New York; and Hampshire County, Massachusetts.

Calumet County is east of Lake Winnebago in the dairy country of northeastern Wisconsin. The rolling plains are the gift of the glaciers, which plastered the countryside with deep deposits of lime-rich debris when they melted. The soil is sweet and fertile, and most of the land is gentle enough for easy cultivation. Even the steeper slopes, where the snout of the glacier lay stalled for centuries, can be used for hay crops or pastures, which protect the soil from serious erosion.

---

* Chapter 10 in *The Land That Feeds Us* (New York: W. W. Norton & Company, 1991).

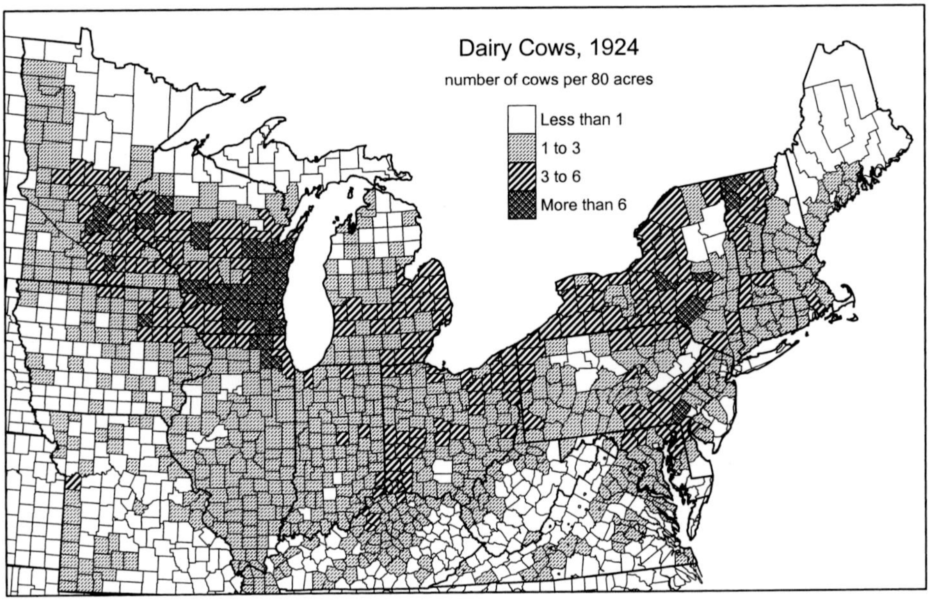

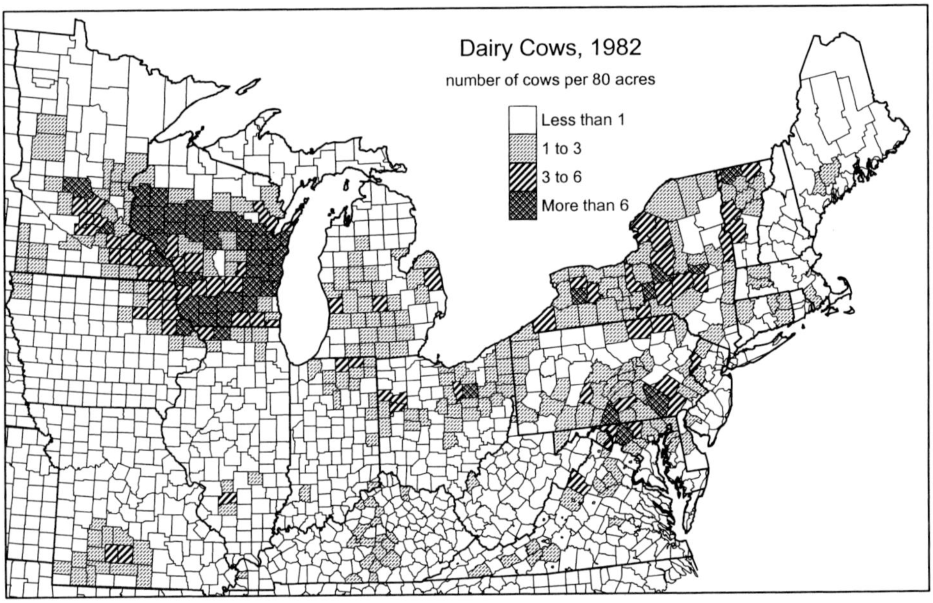

FIG. 12.1. Dairy cows per eighty acres in 1924 (top) and 1982 (bottom). In 1924 nearly every farm had a family cow, but by 1982 dairy cattle had become concentrated in a few favored areas.

Fences have almost disappeared from the landscape, apart from a few forlorn relics, because it is more efficient to feed the cows at the farmstead than to turn them out to graze. The farmer who needs to enclose a field temporarily will "slap a hot wire around it." A single strand of electrified wire attached to white insulators on slender steel posts may not look too secure, but cows develop a healthy respect for it once they have satisfied their natural curiosity by licking one of the insulators with their damp tongues.

The farmsteads are handsome. Cylindrical, concrete silos tower over red barns with huge haylofts above solid masonry ground floors. Numerous windows admit light and air to the lower level where the cows are milked. Larger farms have several silos, including one or more gas-tight models of blue steel. Quite a few farms have their names emblazoned in bold white letters on the barn above the name of the owner. The substantial, two-story farmhouses are set on lawns that would be the envy of most suburban homeowners. Most farmsteads are graced by colorful flower beds, but few have kitchen gardens because it is cheaper to buy vegetables at the supermarket.

In 1982, when I visited Calumet County, Tim Meyer, the energetic county agricultural agent, told me:

> "The hard-working German-Catholic farmers of this county have big families with lots of kids. Some of the kids drift away, but many of them stay on the farm. They really are tied to the farm. Going to town means Chilton [population 2,965] or Kiel [3,083], and the city means Fond du Lac [35,863]. When I was teaching vocational agriculture in high school, I took a bunch of them to a convention. Some of them had never stayed in a motel, and they didn't even know how to order a meal in a restaurant or pay for it."

Tim suggested that I visit Ralph Steiner:

> "Ralph is a good family farmer. He isn't hung up on appearances, and he doesn't feel compelled to spend money on new equipment—in fact, he makes a lot of his own. He's extremely innovative and is constantly trying out new ideas. He's a very smart farmer, and his boys were some of the brightest students I had when I was teaching high school."

Ralph's farm is on the gently rolling glacial plain four miles southwest of New Holstein. A bright plot of flowers adorns the lawn in front of the two-story, white, clapboard farmhouse. Next to the flower bed is a large, white sign with the oval red, white, and blue logo of the Lake-to-Lake dairy company; bold black letters at the bottom proudly proclaim "Ralph Steiner Family."

A gravel drive leads between the house and a one-story, red, wooden workshop on the right to the big, red barn, which is set one-on to the highway. An earthen ramp gives access to the upper level of the barn, whose lower level has been extended by a cinder block addition. A concrete silo and two blue, metal silos are near the barn, and a fourth silo stands in one of the securely fenced "loafing yards" for dairy cows. Each loafing yard has its own metal shed. Downslope is a concrete pond for holding liquid manure until it can be spread on the fields.

Ralph, fifty, said:

> "I think my grandparents on both sides came here straight from Germany. I took over the farm from my dad when he was sixty-five and I was twenty-five. He was born here. Alice and I have four boys and two girls. Bob wants to take over the farm. I think I may let him and start working for him instead of paying him to work for me. I was twenty-five when I started farming, and he should have the same chance I had. I don't want to make him wait until he's forty before he starts on his own."
>
> "I don't buy new machines; they cost too much. I love making my own machinery. I have made a tank spreader for liquid manure, a chopper wagon, tractor loaders, and an elevator. We do most of our own building and all of our own concrete work. We built the concrete manure pond. We made our own forms of plywood and two-by-fours and had to move them seventy-five times when we were pouring concrete. I plan to pave all of the loafing yards. I bought a concrete mixer that operates on the back of a tractor. It was a terrible headache until I installed a hydraulic motor, but ever since then it has worked like a charm."

The big, old, red barn has been greatly remodeled. The loft was used to store loose hay until 1948, when they started blowing dry, chopped hay into it, but barn fires from spontaneous combustion became so common that Ralph switched to bales. He no longer makes any hay, and he uses the loft to store baled straw for bedding. The ground floor of the barn has been expanded to twice its original size and now has forty-four stanchions.

Ralph built the concrete silo eighteen years ago to replace an old stave silo. It holds corn silage to feed the milking cows in winter. He built the second thirteen years ago to hold silage for the heifers and dry cows that are in a separate loafing yard with its own shed. Three years ago, he bought the twenty-by-thirty-foot, blue, metal silo for high-moisture corn when he started producing more than he needed to fill the silos. The metal silos are easy to unload into feeding troughs in the loafing yard, the cows love the feed, and there is no waste.

Ralph Steiner owns 120 acres of land and rents forty more from his brother. He had eighty acres of corn, sixty acres of alfalfa, sixteen acres of oats, and was experimenting with four acres of soybeans. He grows corn for four years, or until weeds get too bad, and then seeds the field to alfalfa. At the end of September, he cuts enough corn to fill the concrete silos. He gets twelve to fifteen tons to the acre, and it usually takes about forty acres to fill the silos. In mid-October, he combines the rest of his corn when its moisture content gets down to around twenty percent and stores it in the small, blue metal silo.

After corn, he seeds alfalfa with oats or Canadian field peas. He tried some direct seeding without a nurse crop, but it had to be done early because hot weather really hurts the young alfalfa plants. Planting peas is more efficient. The peas grow four or five feet tall, shade the alfalfa, and make good haylage. The cows like them better than oats, too. Ralph cuts his first crop of hay around the first of June, dries it in the field for a day or so, and then blows it into the big, metal silo. The second cutting takes place around the beginning of July. He gets about five tons of haylage in a good year. Some of the lower and wetter fields can only stay in alfalfa a couple of years, but the higher fields stay longer. Then he goes back to corn again.

The entire farm operation is based on the hay crop. You can maintain a cow on roughage, but she needs protein for growth, milk production, and the development of her unborn calf. Haylage is good roughage, and alfalfa has twice the protein content of corn. Corn silage also provides roughage, but you have to feed a cow additional protein when you are using it, and the high-producing cows need large amounts of protein. High-moisture corn is a good source of easily digestible energy to complement the protein in alfalfa. It is excellent for high-producing cows, but dry cows and heifers get too fat on it.

Ralph said:

> "I feed the cows their protein supplement when we are milking them. We milk twice a day, at 6:30 a.m. and at 6:00 p.m., after supper. It takes about an hour and a half in the morning, but only an hour and a quarter in the evening because the time between milkings is shorter. I start milking the cows when they are two years old, and four 305-day lactations is pretty good. After that the cows go to make hamburger. I average about 16,000 pounds of milk per lactation.
>
> "I milk more cows and buy more feed than I did when I started farming. I started with twenty-four cows, and now I'm milking forty-seven. The way to expand is to add more cows and to buy more feed, not to get more land. You can buy feed and still make money milking cows. Many years it would be cheaper to buy corn than to grow it yourself, especially when you figure in the cost of the machinery."

I went back to the farm after supper to watch the milking operation. Ralph has had cartilage operations on both knees, and Alice usually has to help him out with the milking, but this evening Barbara, sixteen, and Janet, thirteen, were on duty. They both knew exactly what they were doing. Janet opened the door to the loafing yard and the cows came jostling in. They found stanchions, where Barbara tied them up. Janet walked down the line with a big bucket of protein pellets under her arm. She put two cups of pellets in front of each cow. Barbara pushed a wagon of soapy water down the center aisle, and Janet washed each cow's udder. Barbara attached the milking cups to each cow's teats and coupled them to the overhead vacuum line. Janet came back with a wheelbarrow of high-moisture corn and placed two scoops in front of each cow, three in front of the high producers. Barbara removed the cups from each cow after she had been milked, gave her a slap on the rump, and sent her back out to the loafing yard.

Ralph kept moving cows from one side of the barn to the other because only one side has a milking line. One old cow –"a problem cow," Ralph muttered—kept pawing at the milking cups with her hoof and trying to scrape them off. Ralph had to sit on an overturned bucket beside her while she was being milked to make sure she behaved herself.

Milking cows is an important source of farm income in upstate New York, just as it is in Wisconsin (Figure 12.1). Farmers in New York, in fact, depend on their milk checks for an even greater share of their income than farmers in Wisconsin, but they produce and sell less milk per square mile because nature has endowed their area less generously.

Upstate New York, like eastern Wisconsin, was remodeled by glaciers, but in New York the glaciers were bulldozing their way uphill into the uplands of Appalachia. They scraped smooth the hillsides and gouged out the valleys, some of which later filled with water to become the Finger Lakes. The ice, when it melted, plastered the countryside with stony debris, and it left a legacy of thin, sour, stony soils that require careful management.

Upstate New York is downwind from the Great Lakes, which cool and humidify the air masses that cross them. Gray, overcast days are common, and precipitation is abundant, even excessive, but it is the kind of gentle steady rain that soaks the soil without seriously eroding it.. Spring comes a week or two later in New York than it does in Wisconsin, the growing season is several weeks shorter, and the summers are a few degrees cooler. These differences may seem slight, as indeed they are, but, near the margin, even slight differences can be critical. It is possible to grow corn in upstate New York, but hay is a safer and more reliable crop. A quarter of the farmland is wooded, and two-thirds of the cleared land is used for hay and pasture.

Upstate New York was at the forefront of American agriculture between 1800 and 1850. Farmers made fortunes growing wheat and could afford the luxury of experimenting with new ideas. The area became a leading center of agricultural innovation, and prosperous farmers and small town merchants built fine mansions for themselves. The handsome small towns and pleasant rural areas of upstate New York have been discovered by affluent hordes from New York, Buffalo, Rochester, Albany, and other large cities within easy driving distance. Picturesque, old country inns have been modernized, refurbished, and overpriced, and nearly every fine, old house in nearly every fine, old village seems to have been turned into a fine, old antique shoppe.

After 1850, the center of wheat farming shifted to the Midwest, and most farmers in upstate New York shifted to dairying. The standard dairy barn had room for only about twenty cows. The expansion of dairy herds in recent years has forced farmers to enlarge their old barns or to build new ones. The newest barns are one-story structures with housing for cattle. They are cheaper to build and cheaper to insure because hay fires are a constant threat in barns with lofts.

A modern dairy farmstead will also have a shed for machinery and separate sheds and yards for heifers and dry cows.

Dave Timmerman has a splendid, old red barn on his farm on the steep hillside south of Oneida Lake in Madison County, New York. The barn is crowned by two large, louvered cupolas and flanked by two tall, concrete silos. Dave, forty-five, was too busy to visit with me, but he did take time to tell me that his great-great grandfather had bought the 190-acre farm in 1855 and built the barn in 1875. Originally, the barn had stanchions for milking thirty cows, but Dave has enlarged it to hold sixty. Dave told me:

> "Once this was one of the largest farms in the county, but now it's one of the smallest. My grandfather had six hired men, but today it's just barely big enough to support a single family, and I am going to have to get bigger if my son, who is eleven, wants to join me when he finishes school. This is no life for anyone who doesn't want to farm, but I love it. When I am up there on the hill driving a tractor, and I can look out and see Oneida Lake glittering in the distance, I wouldn't trade places with anyone in the entire world. I've got to be careful where I'm looking, though, because some parts of the farm are so steep that it's dangerous to drive the tractor downhill."

I had a surprise when I walked into the county agent's office in Madison County in 1982 because I had never before met a female county agent. As soon as I saw the name of Kathryn E. Brown I realized that she had to be better than good to be successful in a traditionally male job. She was. She grew up on a dairy farm in western New York and she definitely knew her business. She told me:

> "Our big advantage for dairy farming here in upstate New York is growing good forage. We have to buy concentrates, such as soybean meal, but we can raise just about everything else we need to feed dairy cows. You should have at least two and a half tillable acres to feed each cow. We have good growing weather, with mild, moist summers, but some years are a bit too moist.
>
> "Three-quarters of our tillable ground is in alfalfa and timothy grass hay. We cut the first crop of hay around the end of May or as soon as it stops raining and get a ton to ton and a half per acre. Then we cut at six-week intervals and get a ton and a half to two tons from the later cuttings. We can

bale the second cutting, but we chop the first and third for haylage because the weather in early and late summer usually is too wet to make good hay.

After four years of hay, we plow up the field and plant corn. About a quarter of tillable land grows ninety-day corn for silage. We get fifteen to eighteen tons to the acre and up to twenty-five in a good year. Some farmers combine part of their crop as high-moisture corn, but it's cheaper to buy shelled corn than it is to try to grow it here because our summers are fairly cool for corn. High-moisture corn also requires special storage facilities and good management because you have to feed it as soon as you take it out of storage. We grow corn on the same field for three years because we use atrazine to control weeds the first year, and the residue will kill alfalfa. Then we plant oats as a cover crop for hay, although some farms seem to grow oats simply out of tradition. We seed oats around April fifteen and corn around the first of May."

"Pastures are not as important as they used to be, and most farmers use them mainly as exercise areas for their cows. A lot of our pastures have gotten badly infested with thornapples. The low, bushy plants have pretty, white flowers that city people find attractive, but farmers detest them because they are a terrible pest. Cattle eat the fruit and spread the seeds with their droppings, and the plants are almost impossible to eradicate. Some thornapple thickets are so dense that not even a rabbit can get through them."

Kathy suggested that I talk to Harlan W. Jones about his operation. The Jones farmstead, a few miles south of Cazenovia, is as pretty as a picture postcard. All of the buildings except the two-story, white, frame farmhouse are painted brilliant red with neat white trim. They form a north-facing U around a large, tree-shaded lawn that could be used for a putting green. A brand-new machine shed and workshop on the west side faces the farmhouse and a range of low buildings on the east. The south end is closed by a T-shaped barn with windows on the ground floor and a huge gambrel roof. The barn is adorned with a lifesize painting of a black and white Holstein cow beneath "Jonesville" and above "H. W. Jones" in white letters. The stem of the T, which extends southward, is a single-story structure with many windows. It is flanked by two concrete silos and two blue, metal silos.

Harlan said:

> "I like bright red paint. Regular barn paint starts to look three years old before it has even begun to dry. We built the barn in 1970 after the old barn burned. The old barn held forty-eight cows. The new one has seventy-four stanchions. I don't like free-stall housing because it's too dirty. My father said if you can't make a living milking seventy-four cows you'd better quit.
>
> "He bought the 100-acre farm here for taxes in 1935, and he added fifty-five acres later. I bought the farm from him in 1964, and over the years I've added 215 more tillable acres. He's eighty-four years old. He was incapacitated by a stroke a year ago and lives in the former tenant house next door. I'm forty-seven and I have been farming since I was five. My son is twenty, and he isn't sure what he wants to do. My wife helps in the fields when I need her, and she keeps the farmstead spotless. I have a full-time hired man who lives in Cazenovia and a schoolboy who has worked for me part-time for seven years."

Harlan was milking seventy-seven cows, with thirty yearlings and forty-two heifers. His herd average is 18,115 pounds of milk. It was down a bit from a couple of years ago because he was milking two more cows. The average age of the cows was forty-nine months. He sells them for beef as soon as they start having breeding problems. If he can't make money on them, he said, nobody can, and it wouldn't be right to sell them to anyone else. He also sells bred heifers in the spring if he does not need them for replacements. In winter, he keeps all of the cows in the barn, where they are warmer. In summer, those milking under fifty pounds a day are turned out to pasture, but he keeps the high milkers in the barn, where it is cooler.

The milking parlor and the milk room are in the front part of the barn, beneath the hayloft, and the cows are in the one-story southern extension. They are tied to their stanchions with chains, and high-moisture corn and haylage are placed before each one. At milking time, they are led out through the west yard into the milking parlor at the west end of the barn. Each cow gets five pounds of protein concentrate while she is being milked, and then she is led back to her stanchion and tied up again.

The milk room is in the east end of the main barn. Harlan took down from the wall of the milk room a certificate attesting that his farm is above the state average in milk yields, and he showed me a set of figures he penciled on the back. In 1965, he was milking sixty-five cows that produced an average of 15,900 pounds of milk with 575 pounds of butterfat, and in 1980 he was milking seventy-four cows that averaged 19,290 pounds with 703 pounds of butterfat.

I asked him to explain the phenomenal increase, and he said:

> "Good feeding, good breeding, and good management! My father used to say that a good bull is half a herd, but a poor bull is the whole herd because he can ruin you in no time at all. Your average drops for a little while when you start milking more cows, but you should soon get it back up to where it used to be. This really is a small farm, but I have to work 80 to 90 hours a week in summer and 40 to 50 in winter just to make a living. I have enough headaches with 77 cows that I don't want to get any bigger, but I may have to."

When Harlan said his farm was small, he reminded me of Edward Montague's farm in the hills of central Massachusetts, which was truly small. Ed was milking twenty-five cows and just barely managing to hang on by his fingernails when I first visited his farm in 1958. He owned 350 acres, but he had to buy much of his feed because only thirty-one acres were tillable. The rest of his farm was severely cutover hardwood forest or steep and stony pasture that, in his words, was "hardly worth anything. Juniper has ruined all of our pastures, and you can't kill it."

Ed's farm was in the hills west of the broad valley of the Connecticut River. Roger Harrington, the ebullient county agent in Hampshire County, said:

> "The hills and the valley are like night and day. The hills are mostly woodland and stony pasture, but parts of the valley are pretty good farming country. Valley agriculture is mostly specialty crops such as vegetables, potatoes, and tobacco."

Ed's farm was five miles northwest of Northampton by a narrow, winding country road lined with massive stone walls. The handsome, old, two-story, white frame farmhouse sat next to the road. A large, unpainted shed connected it to a weatherbeaten, old hay barn in the back. Ed, fifty-six, told me:

"My mother's people built this house and the old barn in 1813. We installed plumbing in the house in 1913. We cook with oil, but we still burn wood in winter. We built the new barn in 1930. It has twenty-four stanchions, but you need at least thirty-five to making a living.

"The cows get forty pounds of corn silage, fifteen to twenty pounds of grain, and 3 pounds of citrus or sugar-beet pulp every day, winter and summer. I buy citrus pulp from Florida or sugar-beet pulp from Minnesota in hundred-pound bags. The cows get fifteen pounds of hay in winter, but only five to ten pounds from May 'til October, when they're out on pasture. I'm paying three dollars a hundredweight for grain, if you can imagine that! That's where all our money goes. That's why we can't pay our bills. What I'd like to do is to have the government throw out all these price supports so I could buy my grain cheaper.

"I have six acres of corn and twenty-five acres of mowing land. I rent another thirty acres of mowing land and draw hay from there for the winter here. We put up about 2,000 bales of hay each year. We make two cuts, and then let the cows eat the third. We put all of our manure on the corn land and fertilize it pretty heavily. We cut the corn for silage.

"About all the woods are good for is maple syrup. I tap about 600 buckets of maple sap every March and get around 100 gallons of syrup. People come right up to the house and want to buy maple sugar when they see smoke coming from the sugaring house, but most of it is already ordered in advance for five dollars a gallon.

"I do day work a good deal. I get a little road work every now and then. I used to work in the woods full time and did chores morning and evening, before and after work. I still do a bit of hauling for a sawmill. A man would have to be crazy to try to farm here. We don't make enough to pay our bills, only just enough to keep the wolf from the door. I only do it because it's the old home place. If I didn't love it I'd leave."

In the summer of 1982, I went back to visit the Montague farm. As I drove out from Northampton I was curious about what I was going to find. Ed probably had passed away, and I wondered what had happened to his farm. Had another farmer taken it over? Or had the fields been abandoned and allowed to grow up in brush? Or had the land been sold to city people who wanted new country homes on nicely wooded lots?

When I finally turned the last bend I discovered that the farmstead had hardly changed. It had been spruced up a bit with new paint and bright beds of flowers, and the gravel barnyard had been blacktopped. The place looked better maintained than I remembered it, but obviously it was still a working farm. The sign on the mailbox said "Peter and Mary Montague." Now who, I thought to myself, is Peter Montague? None of Ed's sons was named Peter. My knock at the door was answered by a handsome young man with a lean face and a shy smile. "Oh," he said, "I'm Ed's grandson," and I immediately recalled the tow-headed nine year old who had been running around the farm when I was there in 1958. He went on:

> "My grandmother died in 1965, and my grandfather went all to pieces after her death. He sold off all of the cows, then a year later he bought some more, and then he sold those too. He leased out the tillable land, and sold some of the woodland to neighbors when he needed money to pay his bills. He got a job as a night watchman, and he hated it. He finally died in 1977.
>
> "None of his sons was willing to take over the place after he died. My dad tried farming, then trucking. Now he has a job as a janitor, which he likes. I had always wanted the farm, but my grandfather refused to sell it to me because he said I couldn't make a living on it myself, much less raise a family on it. I worked on another farm for a while, and then in 1974 I bought into a partnership with John Serafin, who has the next farm down the road.
>
> "I bought this place from the estate after my grandfather died. He used to say that he would be happy to sell it to any one of his boys for $25,000. Well, I had to pay $60,000 for it when the estate was settled, and I got it that cheap only because it was in the family. I lease the tillable land and the buildings to the partnership. There are just two of us, John and me, and we share fifty/fifty. Sooner or later one of us probably will buy out the other. I suspect it will be me because he's fifty-seven and I'm only thirty-three."

Eight years ago, they were milking twenty cows averaging 12,000 pounds. In 1982, they were milking forty-two averaging 17,600. The cows were all registered, which meant a better market for the surplus cattle. The milking cows were at John's farm because he had a bigger barn, and Peter kept thirty-five heifers and dry cows at his place.

Peter had thirty-one acres of tillable land, John had forty-four, and they rented twenty-five "in little bitty pieces scattered all over the place." They grew forty acres of corn and sixty acres of hay. They cut all the corn for silage and stored it in a pit silo at the other farm. They cut 12,000 bales of hay each year. They made the first cut at the end of May, another six weeks later, and then they pastured or green-chopped it. Green-chop went straight from the field to the feed bunkers. They tested their silage and hay for protein a couple times a year and bought grain accordingly. A cow got a pound of grain for each three pounds of milk she produced. They no longer bought citrus pulp because the price got too high. Peter said:

> "The sugar house went to hell when my grandfather got too old, and he sold off the equipment. I started again two years ago and have made about 100 gallons a year, but I've given most of it away. He liked horses, but I use tractors to haul the tanks of sap in from the woods. I still set buckets but this year I tried some plastic tubes, and they seemed to work pretty well. I make a living by doing half a dozen different things. Our partnership owns a small sawmill, and we sell some lumber. We still burn wood in the farmhouse, and I sell a bit of firewood. I wouldn't think of buying it because it's too expensive, but it doesn't cost me anything when I cut it myself. I opened a gravel pit down along the stream, and I sell gravel to a local gravel company. I guess you could say that the dairy farm buys the groceries, and the gravel pays for maintenance and improvements. My wife, Mary, is a registered nurse, and she works part-time. That helps a lot."

After he had shown me around the place, Peter told me he had to work on his lime spreader. Last fall, he got a very good buy on some ammonium nitrate fertilizer. He had no place to store it, so he had it dumped on a piece of bare ground, and covered it with a sheet of black polyethylene. He does not own a fertilizer spreader, so he had used his lime spreader instead. The fertilizer had corroded the spreader so badly that it had seized up on him, and he had to struggle mightily to force it loose.

I admired his ingenuity and his skill at improvisation, whether at spreading fertilizer or at making a living. I was reminded of an earlier generation, and as I was taking leave of him I could not resist saying, "Peter, your grandfather would have been proud of you."

Change

CHAPTER 13

in the Corn Belt*

The seedbed of American agriculture was the fertile limestone plains of southeastern Pennsylvania, which are the only extensive tract of inherently good farmland on the eastern seaboard of the United States. By happy coincidence, some of the best and most advanced immigrant farmers from Europe settled this superb farmland. The German farmers in southeastern Pennsylvania developed a system of mixed farming that was so sound ecologically and so successful economically that their descendants have practiced it for three centuries in what is still one of the finest farming areas in the United States.

Mixed farming integrates crop and livestock production. The farmers grew corn, small grains (wheat or oats), and clover in regular rotation. They used these crops to put good solid flesh on lean hogs and cattle, and they hauled manure from the animals to their fields to enrich the soil and maintain its fertility. They raised their own hogs, but they bought cattle from drovers.

Professional cattle drovers ranged far and wide through the valleys and hills of western Pennsylvania, Maryland, and Virginia. They bought lean cattle, a few at a time, from individual settlers, and assembled them into large herds that they drove to the sales rings of southeastern Pennsylvania. They sold the animals to local farmers, who took them home to their barnyards to fatten them up for market. ("Fat" has become such a dirty word that nowadays farmers prefer to say that they "feed" lean cattle instead of saying that they fatten them.)

Migrants from southeastern Pennsylvania transplanted their farming system to the Miami Valley of southwestern Ohio, where the Corn Belt was born, and

* Chapter 2 in *The Changing Scale of American Agriculture* (Charlottesville: University of Virginia Press, 2003).

whence it spread westward across the Middle West. The Corn Belt farming system was based on a standard three-year rotation of corn, small grains, and hay. Corn is the quintessential American crop. The small grain that followed corn was winter wheat in the south, oats in the colder areas farther north. Clover was the principal hay crop in the early days, but eventually alfalfa almost completely replaced it.

Clover and alfalfa are hard to establish on bare ground, so farmers planted the hay seed with their small grain, and the grain served as a nurse crop that sheltered the tender young hay plants. Both clover and alfalfa are legumes. They extract nitrogen from the air and store it in nodules on their roots; the nitrogen enriches the soil for the corn crop that begins the next cycle of the rotation.

Farmers modified the standard rotation in response to environmental conditions on individual farms and in individual fields. On better ground, they might extend it by planting two or more crops of corn before they planted the small grain, and on steep or hilly land subject to erosion they might leave their fields in hay for several years before plowing them up to plant corn.

Corn Belt farmers fed most of their crops to livestock. Corn is the world's finest feed for putting good solid flesh on lean hogs and beef cattle. It also commands a good cash market, and farmers were closely attuned to the corn/hog ratio; they sold their corn for cash if the price of corn was higher, but they fed it to hogs if the price of hogs was higher. They took most of their wheat to the mill to be ground into flour, but they fed some to animals. Oats were important feed for their workhorses, and farmers also fed their hay to their animals.

Hogs and beef cattle were the money animals, but most farms also had a milk cow or two, a flock of chickens that had free run of the barnyard, and other domestic animals. The type of livestock fed varied with the size of the farm. Small farms concentrated on hogs, which required but repaid lots of attention. Hogs attained a market weight of "two and a quarter," 225 pounds, in only six months. Farmers had to breed their sows, because they could not afford to pay others to produce baby pigs for them to feed.

Beef cattle need much longer to grow to market weight, and the Corn Belt farmers who fed cattle bought lean "feeder" animals from producers in other areas, mainly from ranchers in the West. After harvest in the fall, farmers figured how much corn they had, subtracted the amount they would need to feed their hogs, and bought enough feeder cattle to use the rest.

As a general rule, the size of farms increased westward in the Corn Belt.

Ohio and Indiana, which were settled first, had small corn/hog farms with relatively few cattle. Improved technology enabled farmers to cope with larger acreages in areas that were settled later. Iowa had larger farms, and many Iowa farmers fed lean feeder cattle from ranching areas in the West for the urban markets of the East, in addition to feeding their own hogs.

The exception was the Grand Prairie of east-central Illinois, which was settled late because its vast level plains were hard to drain. The farms of the Grand Prairie were so large that the farmers were fully occupied with their crops, and they had little time for livestock. They sold all of their crops, and thus they were cash-grain farmers rather than mixed farmers.

Corn Belt farming was completely transformed in the last quarter of the twentieth century. Farmers were caught in a cost/price squeeze. The costs of producing crops and livestock rose much faster than the prices farmers received for their products, and the profit per bushel or per head was sliced thinner and thinner. Farmers had to increase their total volume of production to compensate for their shrinking profits per unit. They had to produce more from each acre they farmed, and they had to farm more acres. They increased their production per acre by using better seeds, more chemicals, and bigger and better machinery, and by specializing in what they have learned they can do best.

The transformation of the Corn Belt actually began in 1933, when hybrid seed corn became available. Farmers have always saved seed from their very best plants to sow in future years, but selective breeding by plant scientists produced varieties that were far more productive. Back in the pre-hybrid days, farmers boasted about corn yields of fifty bushels an acre, and even in the 1950s they marveled at the idea of 100-bushel yields, but today they feel sorry for themselves if they do not get well over 125.

In the old days, farmers and plant breeders occasionally lucked out when a natural genetic mutation in the reproductive process produced a valuable new variety of a crop. Some of our best crop varieties are the results of such mutations. Today, we no longer have to wait and hope because biotechnology enables plant breeders to borrow a useful genetic code from one organism and transfer it to another.

For example, for more than thirty years farmers have used insecticides based on synthetic *Bacillus thuringiensis* (Bt) to suppress caterpillars, beetles, and other pests. In the 1990s, biotechnologists transferred the Bt gene to corn and produced a variety that is toxic to corn borers, which can chew up an entire crop in a few days. In 1995, borers destroyed $285 million worth of corn in Minnesota alone, but now farmers who plant Bt corn no longer have to worry about them.

Biotechnologists have also produced soybean varieties that resist certain herbicides, and in 2000 more than half of our soybeans and one-third of our corn grew from genetically modified seed. Farmers like these new transgenic crops because they greatly reduce the need for synthetic chemical pesticides and are friendlier to the environment.

Not everyone is happy with genetically modified organisms (GMOs). Some people are uneasy about the mysterious science of biotechnology, which works its miracles at submicroscopic levels, and small but intensely passionate and highly vocal groups of activists have played on their fears. The activists have used the news media extremely well, while the biotech companies have smugly assumed that they only have to sell their seeds to farmers and can ignore consumers.

Attacking GMOs is also a socially acceptable way of attacking large corporations and demonstrating anti-Americanism, and some countries have even embargoed imports of GMOs. In April 2000, a careful study by the National Academy of Sciences concluded that GMOs pose no apparent threat to people who eat food made from them, but the arrogant biotech companies may be losing the battle of public opinion. One is forced to wonder how hybrid corn might have fared if it had been developed in today's activist climate.

The development of hybrid corn enabled Corn Belt farmers to double and even triple their yields, but it also forced them to start using a whole arsenal of agricultural chemicals. The soils of the Corn Belt are so inherently fertile that they can produce forty bushels of corn to the acre with no additives but leguminous crops and barnyard manure, but farmers had to start pouring on chemical fertilizers to take full advantage of the potential productivity of hybrid corn.

One of the finest chemical fertilizers is anhydrous ammonia, a dangerous chemical that is a superb source of nitrogen if it is injected into the soil properly. Long rows of gleaming white tanks of anhydrous ammonia are a common springtime sight in dealers' yards at the edge of nearly every small town in the Corn Belt. The dealers, often grain elevator operators, also sell insecticides to control insects and diseases, and they sell herbicides to kill weeds so farmers no longer have to cultivate between the rows of crops to eradicate them.

Increased yields and greater use of agrichemicals have been associated with larger, more sophisticated, and much more expensive machinery. As late as 1950, some people argued vehemently that tractors would never completely replace workhorses, but few contemporary Corn Belt farmers would even know how to start hitching up a team. The modern farmer has three or four tractors, maybe more at $50,000 or more a pop, a combine with a list price well over $200,000, and a total machinery inventory of $500,000 to $1 million or more. You could have equipped an entire farm in 1950 for the price of the electronic monitor on a modern corn planter.

The self-propelled combine harvester has revolutionized corn picking. Much of the crop was picked by hand back in the days of forty-bushel corn, and husky young farm boys liked to boast that they had picked 100 bushels in a day. During and after World War II, mechanical pickers replaced hand picking. They tore the entire ear from the stalk, but the farmer had to shell the kernels from the ears after they had been picked. Combines, which both pick and shell corn, began to replace mechanical pickers in the 1960s. An early four-row combine could pick and shell 100 bushels of corn in less than ten minutes, and now six-, eight-, and even twelve-row models are common.

Combines have two harvesting heads, which a farmer can easily change in half an hour. The farmer harvests soybeans with the revolving reel-type head, then changes to the corn head, which has large, blunt, metal prongs that nose between the rows of corn. The loose beans or corn are stored in a large bin at the top of the combine. They are unloaded through a tubular spout that swings out to one side over the metal tractor-drawn wagon or truck that hauls them back to the farmstead.

Traditional wooden corncribs with open slatted sides cannot hold loose shelled corn, and they are disappearing rapidly. Combines have forced farmers to invest in new cylindrical grain storage bins of gleaming corrugated metal. Most farms have nests of several bins, with a tall metal "leg" housing a bucket elevator towering above them. The elevator lifts loose corn and soybeans to the distributor head, which directs them through a spider-like aerial network of six-inch metal tubes to the desired bin. Most bins have some kind of drying system, because grain in an enclosed bin must be dried to keep it from becoming moldy.

Farmers in the Corn Belt have started to put global positioning systems (GPS) on their combines, tractors, and other machinery to enable them to farm more precisely. GPS, which were developed by the military but have been made available to the public, use orbiting satellites to pinpoint locations on the earth's surface to within a few feet.

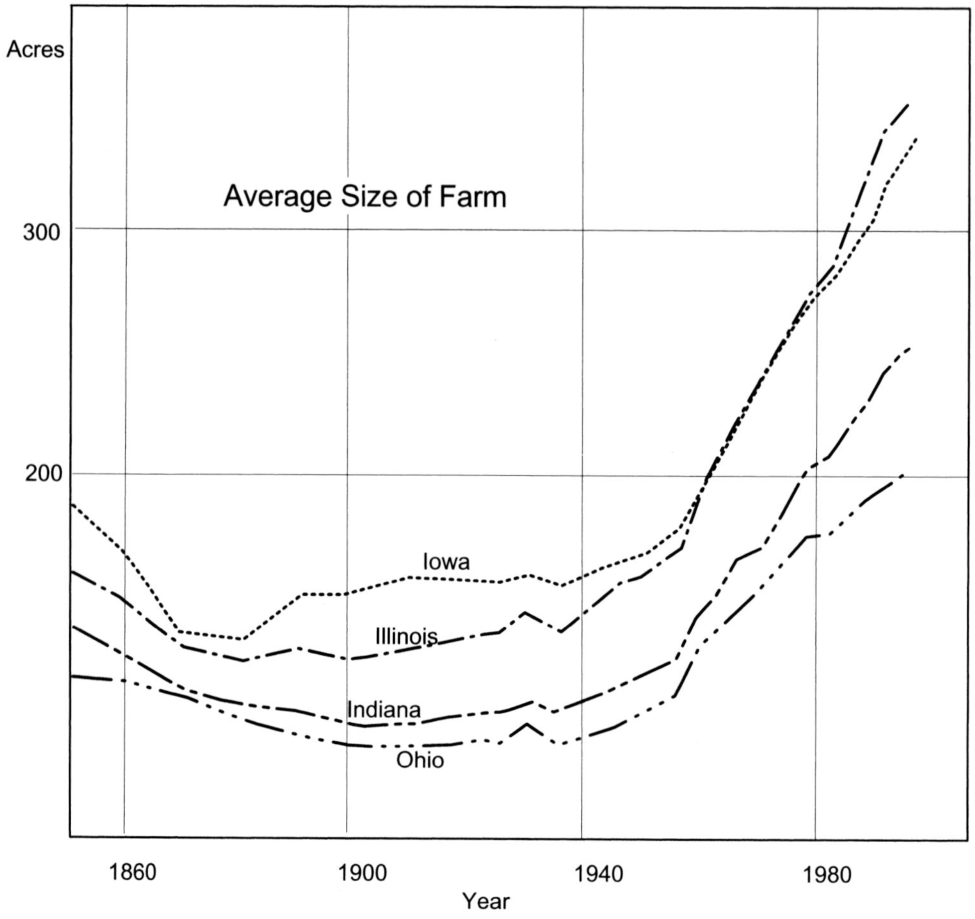

FIG. 13.1. The average size of a farm in four Corn Belt states.

In the past, farmers had to treat each field as though it were completely uniform throughout, even though they knew better, because they were not able to adjust their equipment to compensate for variations in soil fertility, drainage, slope, and other factors that affect their yields. They blanketed the entire field with uniform applications of fertilizers and pesticides, too much in some parts of the field, not enough in others.

The development of variable-rate applicators now enables farmers to place chemicals precisely where they are needed, just the right amounts in just the right places. They no longer waste money putting chemicals where they are not necessary, they make more money by increasing their yields, and they avoid any possible undesirable environmental effects of excessive application.

Combines now have digital yield monitors that measure the force of the harvested grain beating against a fixed plate to estimate the yield of grain at one-second intervals. The GPS pinpoints these measurements to within a distance of a few feet to produce a detailed map of the field. The farmer compares this yield map with a detailed soil map of the field to determine which parts need more fertilizer next year, and how much more.

When the crop is growing, the farmer hires a pilot to fly across the field and take infrared aerial photographs, which clearly show healthy plants and weeds. The GPS on the variable-rate pesticide applicator uses this information to spray herbicides on the weedy patches and avoid wasting it on healthy plants that do not need it.

The integration of GPS, variable-rate applicators, yield monitors, and infrared aerial photographs has enabled Corn Belt farmers to tailor their operations much more precisely, but this new technology does not come cheap, and it takes a while for the farmer to learn how to use it effectively. The expense must be spread over such a large acreage that smaller farmers simply cannot afford it, and even some of those who can afford it are not sure that it will ever pay for itself.

The average size of farms in the four Corn Belt states has been escalating furiously since 1950, after having been remarkably stable for nearly a century (Fig. 13.1). Every farmer in the Corn Belt needs to increase the size of his farm, but the intense competition for land has driven the price so high that few farmers can afford to buy it, and most have enlarged their farms by renting land rather than buying it. The farmer who owns part of the land he farms and rents the rest is called a part owner, and part ownership has been the preferred strategy for farm enlargement in the Corn Belt since World War II. Part owners farmed less than one-quarter of the farmland in the Corn Belt in 1949 but nearly two-thirds in 1997, and the average size of part-owner farms shot from 205 acres to 545 acres. Part ownership has increased at the expense of tenants and full owners. Tenant farmers who own no land at all are increasingly rare, and most probably are sons or nephews who expect to buy or inherit the farm when an elderly parent or relative dies.

A modern family farm has become a highly specialized business, with a greater capital investment and greater gross sales than most of the businesses on Main Street. It must sell at least $250,000 worth of farm products each year to provide an acceptable living for a contemporary American family, and many farmers say that even this number is far too low, because they must pay for seed, chemicals, machinery, fuel, taxes, interest, insurance, repairs, depreciation, and all other farm expenses before they can spend any money on themselves. As one farmer said to me, "An awful lot of money passed through this place last year, but not much of it stuck."

Many Corn Belt farmers supplement their farm income appreciably by taking jobs off the farm, and the percentage who do so is increasing. In 1950, only 16.5 percent worked off their farms 200 days or more (which is tantamount to a full-time job), but the figure had risen to 36.8 percent by 1997, and it was even higher in the more heavily urbanized eastern part of the region, where off-farm jobs are more plentiful.

The percentage of Corn Belt farm income derived from the sale of livestock has decreased with erratic regularity since World War II. The sharpest drop was around 1970, when beef cattle feeding relocated from the western Corn Belt to the southern High Plains of Kansas and Texas. The decline in livestock production has reduced the supply of fed animals for processing plants, and some plants have had to shut down.

Many Corn Belt farmers have had difficulty adjusting to the way American agriculture is changing. Their forebears settled in an area that has been singularly blessed by nature. The soils are so productive and the climate is so propitious that it was almost impossible to fail as a farmer if you were to work long and hard and to live frugally. The person who could do nothing else could always make a living as a farmer, or so it seemed.

The old ways are no longer good enough, because the scale and organization of American agriculture have changed dramatically since the 1970s. Today,

Corn Belt farmers must compete in a global economy, and they have to risk huge investments to stay in business. Successful modern farming requires new thought patterns, but some conservative farmers have tried to mobilize their political muscle and pass laws to thwart change instead of trying to learn how to cope with it.

A modern cash-grain farm in the Corn Belt needs 800 to 900 acres of cropland to produce $250,000 worth of corn and soybeans, and that is hardly enough to keep the farmer busy. Bill McCue, who works full-time as a Certified Public Accountant in Minneapolis and has a part-time farm near Belle Plaine, Minnesota, told me in August 1998 that 700 acres of corn and soybeans was about all he could handle because he could farm only on weekends.

Eight hundred acres, though, is the equivalent of more than five 160-acre homesteads, which means that four of every five original homesteads are redundant to the needs of contemporary agriculture. The Corn Belt landscape is littered with abandoned farmsteads and undersized last-generation ma-and-pa farms (Fig. 13.2). The land is paid for, the old folks can survive on the accumulated capital, and some politicians erroneously assume that such places are "family farms," but the children of the family would not be foolhardy enough to accept a lifetime of arduous physical labor with almost no income on such "farms." Neither would the politicians.

Ken Mather and Doug Magnus are Corn Belt farmers who have shifted from mixed crop-and-livestock farming to specialized cash-grain farming of corn and soybeans.

Ken Mather

Ken Mather has actively participated in the transformation of Corn Belt agriculture. In 1851, his great-great-grandfather, Samuel Mather, rode west from Ohio on horseback and bought a 250-acre farm a mile east of Springdale, Iowa, which is fourteen miles east of Iowa City (Fig. 13.3). The next year, he and his wife loaded their fourteen children and everything they owned into two ox-drawn covered wagons and moved to the new farm, which has been farmed by Mathers ever since.

In 1910, his oldest son, Samuel Jr., bought another 140 acres south of Springdale. During the Great Depression, an insurance company foreclosed on a loan it had made to him, and it held title to the farm for ten years before he was

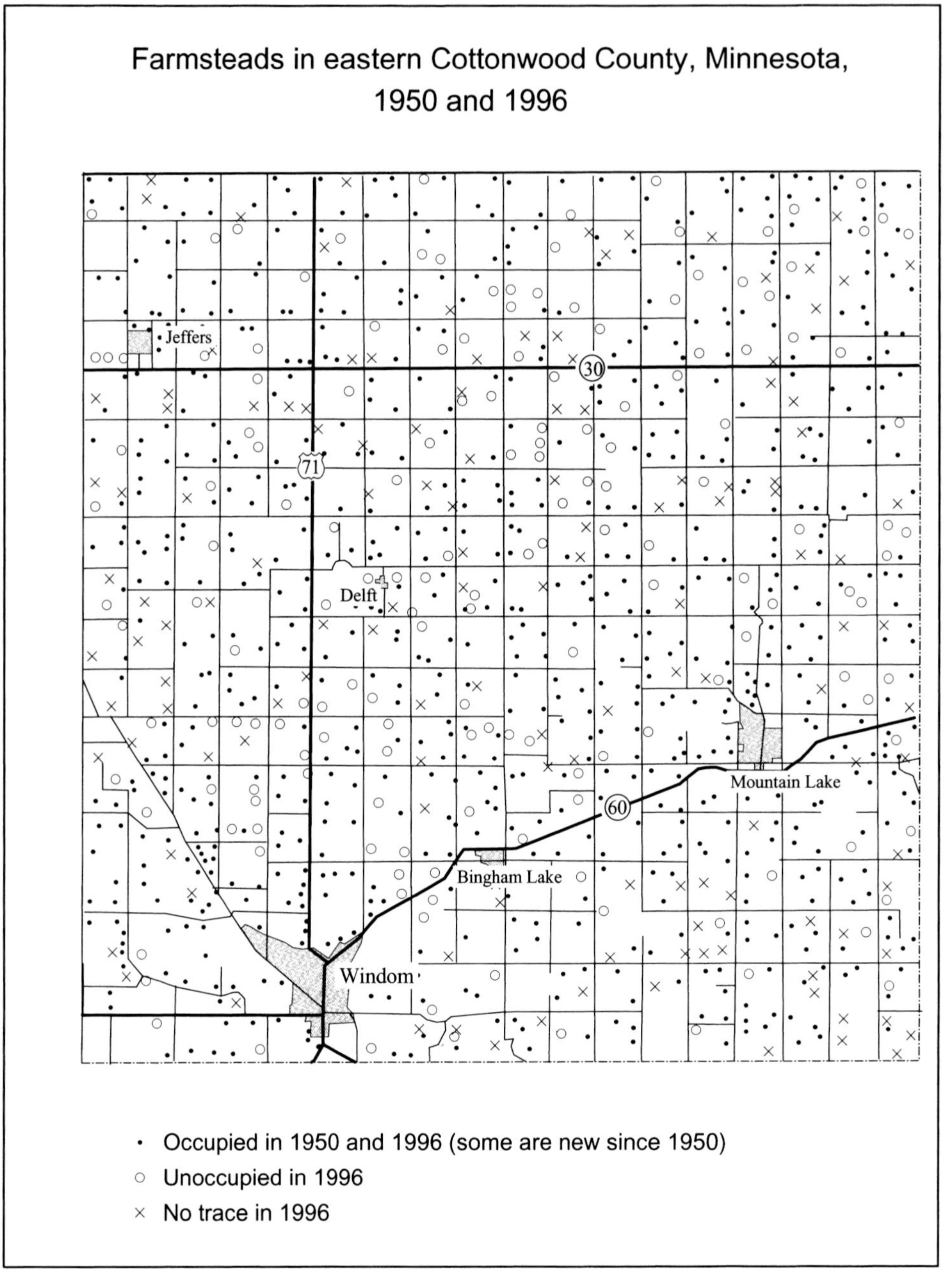

FIG. 13.2. Farmsteads in eastern Cottonwood County, Minnesota between 1950 and 1996. (1950 data are from aerial photographs and topographic quadrangles; 1996 data are from field survey).

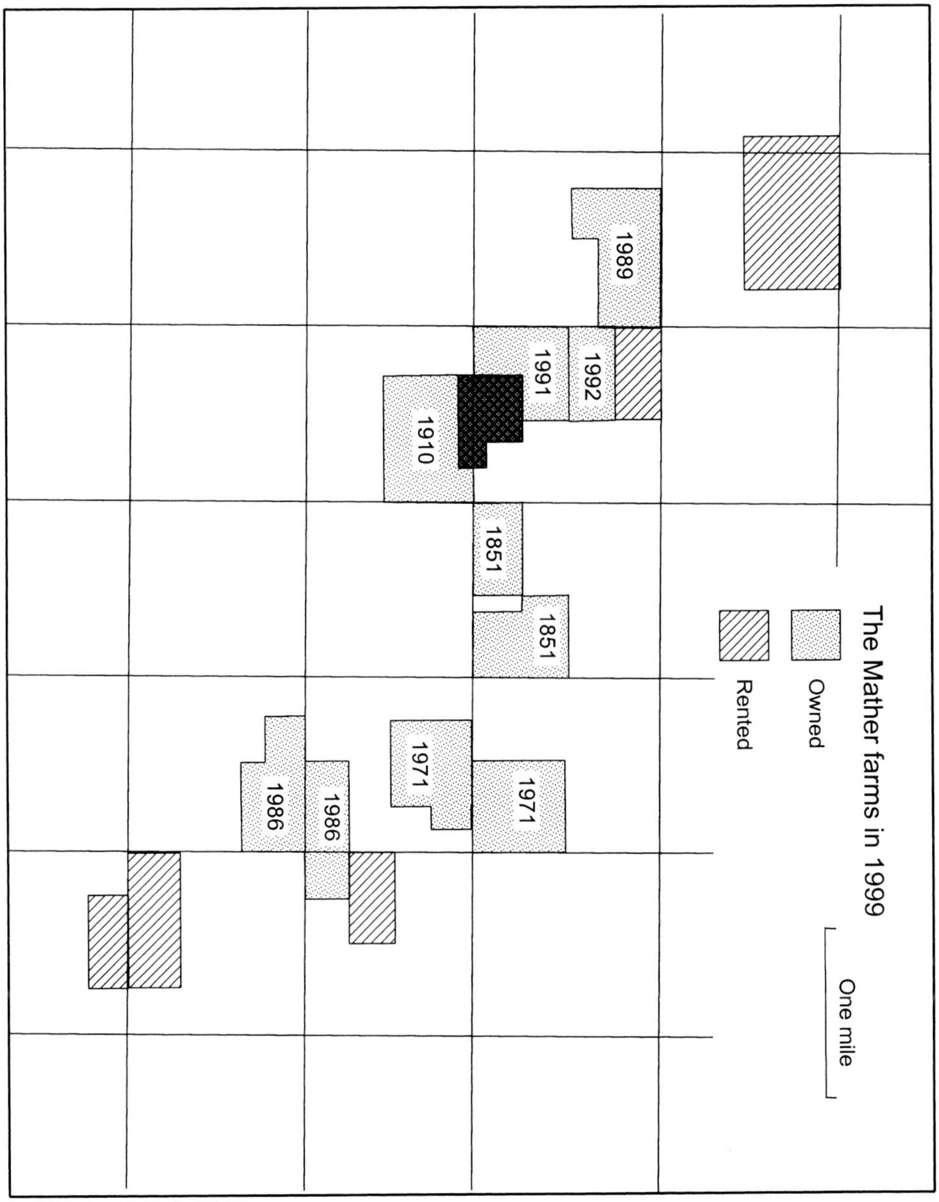

FIG. 13.3. The Mather farms in 1999, showing the date each farm was bought and the farms that were rented in 1999.

able to pay off the loan and reclaim the farm. His son, Anders, rented the farm from his mother after his father died, and he bought it from her estate after she died in 1938.

Anders Mather was an excellent judge of beef cattle. At first he bought feeder cattle from a local dealer, but by the early 1940s he had decided he could do better, so he began making regular trips to ranching areas in the West to buy feeders. His neighbors admired what they saw in his feedlots, and they paid him to buy feeders for them too. By 1950, he was taking orders for feeder cattle from farmers all over eastern Iowa.

Anders bought cattle for fall delivery after farmers had harvested their crops. For his own farm each year he bought 300 calves at 400 to 450 pounds and 200 yearlings at 650 to 700. At first, he fed them alfalfa hay to help them adjust to the full concentrated feed of corn, with a protein supplement of cottonseed meal, which was shipped from the South by the boxcar load. The yearlings reached a market weight of 1,200 pounds in nine months, and the calves took twelve to fourteen months to reach their market weight of 1,000 pounds. Anders always had a couple hundred head of hogs to clean up the fields after harvest and to clean up what the cattle wasted in the feedlot, but hogs were always incidental to cattle.

In 1960, when he was sixty-eight, Anders began renting the farm to his grandson, Ken, who was born in 1940. Ken had gone to the University of Iowa for a year but hated it; he wanted to farm. He really should have gone to Iowa State, the ag school, but his grandfather would not let him, because Iowa State never had a winning football or basketball team. Ken's father, Andy, who had always farmed rented land, bought the farm from the estate for $350 an acre after Anders died in 1964. He continued to rent it to Ken until 1980, when Ken bought it from him for $525 an acre. Ken remembers:

> "In 1960, when I married Jan and started farming, all I owned was $20,000 worth of machinery. Back in those days it was all four- and five-year rotations, corn, corn, oats, and a year or two of meadow. You fed it all. You ran it through the livestock, and nobody even thought of selling grain. Nobody grew soybeans. They were a poor farmer's crop. You could count all of the fields of soybeans in the community on one hand and still not use all your fingers."

In the late 1960s, Ken realized that his cattle were not making him any money because a small Iowa farm feedlot could not compete with the huge new feed yards that were being developed in irrigated areas in the West. In 1971, he had a chance to buy a 250-acre farm a mile east for $1,400 an acre, so he liquidated his cattle operation for the down payment on it. He realized that he had to build expensive new facilities or get out of hogs, so he got out of hogs. "My grandfather would turn over in his grave," he said, "if he knew there were no livestock on the place."

Ken was a bona fide cash-grain farmer when I first visited his farm n 1982. He owned 400 acres and rented 600 more for $150 an acre. He was growing more soybeans because they were a good cash crop, and they broke the buildup of corn diseases in the soil when he grew continuous corn. His corn yields were running 150 to 160 bushels per acre and beans were running fifty to fifty-five, but low prices were killing him. All he could do was hope that the price would improve before he went broke.

Ken obviously was fairly concerned. He said:

> "I can't take on any more debt, and I am already pushing my labor supply. I am full-time, and Randy, nineteen, Alan, sixteen, and Mike, fourteen, help me after school. Jan runs the grain drier and does most of the bookwork. We formed a family-held corporation in 1976 for tax purposes. If he hadn't incorporated, we'd have a desk full of tax receipts instead of a farm."

Things were looking better when I went back to visit Ken in the fall of 1983:

> "I'm looking to get bigger now, because Randy is starting to get interested in farming. I'll use profits to expand. A year ago I didn't expect to, and maybe I'll change my mind next year, but right now looks like a pretty good time to buy land. The price has fallen off the table.
>
> "The prices are about $1,000 an acre lower than they were last year, and when it gets down to these prices, you can buy it to make money. I'm ready to buy if I can find land I like at a reasonable price."

In 1986, Ken helped Randy buy a 150-acre farm. In 1958, his grandfather had looked at this farm and declared that it was not worth the asking price of

$275 an acre, and a doctor bought it as an investment. In 1978, the doctor sold it for $2,500 an acre, which Ken said was way too much, and the guy that bought it went broke. In April 1986, Ken bought it for $1,350 an acre.

When Alan graduated, they bought another farm for $1,100, and both boys have bought additional farms in the $1,400 to $1,500 an acre price range. They also rent land, and in 1999 they grew about 1,250 acres of corn and 750 of soybeans. They plant corn the last week of April and the first week of May. Alan operates the planter, Randy operates the sprayer, "and I'm the gopher," Ken said. "I spend all day going for this or going for that."

They start to harvest around September 20 and hope to finish by November 10. The boys have bought a large tractor-trailer grain truck that can haul grain in from distant fields faster than farm tractors pulling wagons. At the farm they have a drier that can handle the grain as fast as they can haul it in and six large metal storage bins with a total capacity of 12,000 bushels. "Right now the boys are using their labor and equipment to the max," Ken said. "They could take on a little bit more land, but not much. They would like to buy land if they could find some for sale, but it's in pretty tight hands right now."

Ken is uneasy about the corporate takeover of American agriculture and the concentration of control in fewer hands. He said:

> "One company's chemicals won't work with another company's seeds, so you're trapped into buying the whole package from a single company at the price they set. Corporations don't like the way the price of crops fluctuates, five-dollar corn one year, dollar corn the next. They want to squeeze the price as close to the cost of production as they can get it."

Doug Magnus

Doug Magnus, who grows 1,400 acres of corn and soybeans near Slayton in far southeastern Minnesota, has helped to globalize the American soybean industry. I first met him in 1973 when I gave my annual lecture at the South Dakota State Geography Convention. He was then a twenty-two-year-old majoring in agronomy. He had discovered that geographers have so much more fun than agronomists that he was minoring in geography, and we became good friends.

Doug had attended South Dakota State for two years, then was called into the army for two years, including a tour in Vietnam, farmed for a year with his father, Clarence, and decided to return to SDSU to finish his degree. Clarence thought it was a mistake, but Brenda, Doug's wife, who is trained as a medical technician, strongly supported him by taking a factory assembly-line job. "I still couldn't have done it without the GI Bill," Doug said, "and it probably cost me a lot of money, because farm prices were exceptionally good when I was in school."

In 1976, he graduated and started farming by renting a 160-acre farm from Clarence, who owned 440 acres. Both of them also rented land from others, and by 1983 they were farming 1,400 acres, alternating corn and soybeans (Fig. 13.4). Doug said:

> "The price of soybeans normally should be two to two and a half times the price of corn. If it drops lower we plant more corn, and if it gets higher we plant more beans. We are really stretched for planting in spring and harvest in fall, and we couldn't handle any more land. Even now my retired uncle helps us out, and so does Brenda."

Doug and Clarence are big on John Deere equipment, which is painted green, so I was surprised to see on Doug's farm an enormous red six-bottom plow with huge moldboard blades. "Well," Doug said, "you wouldn't know it wasn't a John Deere if I painted it green; the real reason I have it is that Brenda likes it better than a John Deere." Brenda did all of the plowing, 500 acres a year. She also chopped silage, hauled grain in from the fields, chopped stalks, drove the disk cultivator, and drove the truck that hauled grain from the farm to the elevator.

Doug and Clarence sold all of their beans and most of their corn, but they like beef cattle, and they fed out 500 head a year. They bought 500-pound feeders from ranchers at auctions in South Dakota, fed them corn for ten months, and sold them to local packers at 1,150 pounds. In 1983, Doug was also feeding 600 hogs a year. In 1980, he had bought a "bare quarter" (160 acres of land with no buildings) for $2,100 at eight percent interest, and he fed hogs under contract to meet his payments on it. Buying that bare quarter was a mistake; Doug paid far too much for it, and he was relieved when he decided to forfeit it. He was equally relieved to be able to get rid of the hogs.

The early 1980s were tough for farmers. Prices were terrible, and the papers were full of stories about farmers who had lost their farms. Every time I

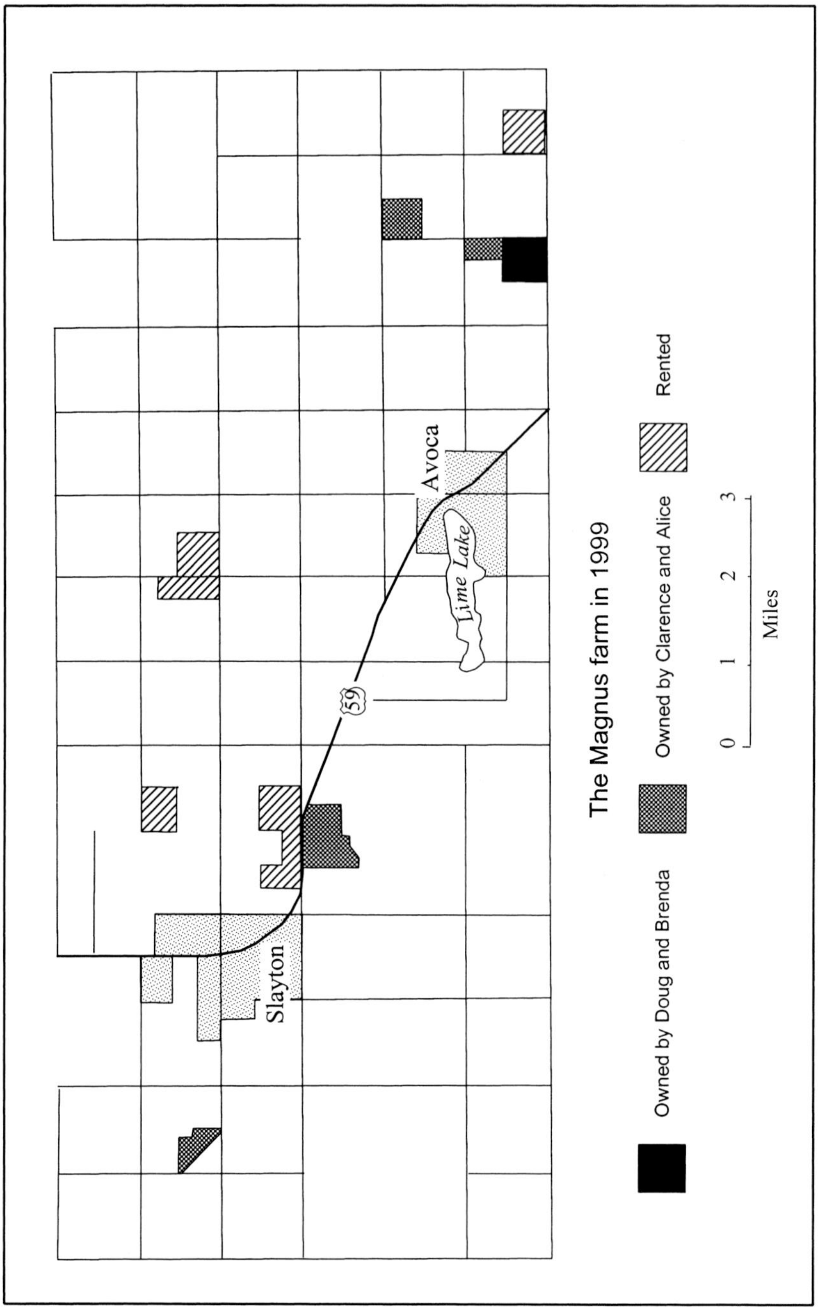

FIG. 13.4. The Magnus farm in 1999, showing parcels owned by Clarence and Alice, parcels owned by Doug and Brenda, and rented parcels.

talked to Doug I was worried until he reassured me that he was still in business. "Somebody is going to farm the land," he said, "even though they are not making any money on it. They've just got to hope that a good year is going to come along and pull them out."

Doug is outgoing and articulate and he has always been involved in public affairs. He started on the boards of local co-ops. In 1988, he was elected to the state soybean council, and he served three terms as chair. In 1992, the governor of Minnesota appointed him to the state Environmental Quality Board, where he served two four-year terms. He is also a consultant for a major farm machinery manufacturing company.

In 1993, the U.S. secretary of agriculture appointed Doug to the United Soybean Board, which is responsible for using check-off funds to promote research, marketing, and the development of new uses for soybeans. In 1994, the board elected him chair of its International Marketing Committee, and in the next six years he led nine trade delegations to see American soybeans in Europe, Russia, Turkey, Southeast Asia ("It was eerie to go back to Vietnam"), China, Brazil, and Argentina.

In 2000, he was elected chairman of the board. One of his first official actions was to welcome to his farm a delegation of twenty feed manufacturers from seven European countries, even though he was properly concerned that they might bring hoof-and-mouth disease with them. He wanted to show them how we grow and handle soybeans in an effort to allay their fears about GMOs.

In 2002, he ran for and was elected to a seat in the Minnesota House of Representatives. I asked Doug how he could find the time for all of his outside activities, and he said:

> "I figure it takes me about three and a half months of full six-day weeks to grow 1,400 acres of corn and soybeans. I need forty-five to fifty days in spring to plant and spray, forty days starting around the first of October to harvest, and ten to fifteen days to haul the grain to the elevator. I can fit in other things around my busy seasons.
>
> "I talked to Dad and Brenda before I made any commitment, and both of them strongly support me. Dad is semiretired now. Brenda works full-time as a medical technician in Slayton, but she loves to help on the farm. When she gets home from work she can hardly wait to change her clothes and get out onto that tractor. Both she and Dad are here to help me when I need them."

As an agricultural leader, Doug is concerned about the number of farmers who are willing to sit around in the coffee shop and complain that it's all "their" fault, and are not willing to recognize the necessity of change. Farmers must shift from production-based thinking to consumer-based thinking, he said.

> "We must give consumers what they want, instead of expecting them to take what we give them.
>
> "We are going to have to figure out how to add value to the commodities we produce, because bulk commodities, such as corn and beans, produce low returns. We are going to have to learn to partner with the companies that add value, even though it is going to cost us some of our independence. I want to sell at least 40 percent of my crops by contract for future delivery at a guaranteed price, because I want the premium price I will get as a preferred supplier/producer."

Doug Magnus and Ken Mather both reflect the shift in the Corn Belt, the nation's agricultural heartland, from mixed crop-and-livestock farming to cash-grain farming, even though they differ in degree. Ken has gotten rid of cattle completely, but Doug still has a sentimental attachment to them, and he lives closer to the ranch areas from which he buys feeder cattle. Doug would like to contract for the sale of part of his crops, but Ken is leery of contracts. They both specialize in growing corn and soybeans for sale, like most farmers in the contemporary Corn Belt.

CODA

John Fraser Hart's fascination with the land, its people, and its structures has sustained him on a remarkable career that has spanned six decades of American geography. The essays in this volume are but a sample of what he has written, but even this small collection possesses a diversity that is comparatively rare in the academic world. Each of the essays stands by itself, yet as a collection they suggest something more that invites a brief commentary. What might have guided one person to pursue such a variety of topics? What interests compelled him to revisit the scenes of his earlier research to update the results over time?

Like all scholars who publish frequently and whose writings are widely known, Fraser Hart has given many opportunities to reviewers and potential critics. One geographer who often took exception to at least some of what he gleaned from the works of others was Terry G. Jordan (1938–2003). In a review of Hart's *The Look of the Land*, Jordan presented a list of complaints. The book was a "shotgun polygamous marriage of rural settlement geography, economic geography, recreational geography, and several other ingredients which defy classification. Would it not have been better for Fraser Hart to confess in the preface that . . . it is impossible to fit these diverse and divergent interests into a coherent system?" "Don't ask us to believe that there is method in your delightful madness," Jordan (1975, 599–80) wrote. Jordan was a follower of some of the older settlement classification schemes devised by German geographers, which Hart had mildly poked fun at in the book. It was a classic confrontation between a younger scholar who yearned for greater order and a mature scholar who distrusted orderly systems and overarching frameworks because they usually obscured more than they clarified.

In the 1990s, Hart offered yet another framework for studying the landscape, only this time it was even less explicit than the one Jordan had found wanting (Chapter 2 in this volume). Hart, meanwhile, continued to move between farming practices in the Corn Belt, recreational developments, and urban-fringe land use, returning to the South now and then to learn more about cotton, pine trees, broilers, and house trailers. More work produced not a neat stack of interlocking concepts, but only more observations, which inevitably led to more questions. Was there method in his madness, as Jordan asked?

If a landscape is changing, then those who study it are free to pursue various options. One is to condemn the changes and perhaps even fight to prevent them from taking place, a common strategy among educated elites. Another is to observe the changes as closely as possible and to advance an understanding of why change is taking place by gathering data and talking to the participants. The latter option has been Fraser Hart's trademark approach, and he has followed it consistently. The common thread connecting the disparate themes is social change itself, in all its many facets. Although Fraser Hart does not ignore the past, he accords it few rights by virtue of simple priority, and he is seemingly forever on guard against the comforts of nostalgia. Seeing what is present is made more difficult if one keeps looking for what used to be.

For example, despite his substantial knowledge and appreciation of traditional American farm buildings, it is hard to find evidence that Fraser Hart was much bothered by their disappearance in the late decades of the twentieth century. When buildings that had ceased to be economically useful were replaced by others that did have value for the farmer, Hart began to notice them and describe their role. If farmers stopped cultivating the land and planted pine trees or sold part of the land for recreational home sites, Hart quickly turned to studying such things as well. In all of these cases, as well as in many others, the landscape and the changes on it are primary. Keeping a fixed gaze while avoiding myopia is a trick that few other geographers have managed to do as well as Hart has.

If anyone ever does try to impose a neat order on the range of topics and insights that his research has produced, I imagine that Fraser Hart will be the first to take exception to it. The ordinary landscapes that he has studied are full of surprises, whether one is viewing a scene for the first time or the tenth. Whatever is there demands to be understood, and that is what Fraser Hart has taught us so well over these many years.

REFERENCES

American Farmland Trust. *Saving American Farmland: What Works* (Northampton, Mass.: American Farmland Trust, 1997).

Brueckner, J. K. "Urban Sprawl: Diagnosis and Remedies." *International Regional Science Review*, 23, 2 (2000):160-71.

Claval, Paul. "A *Grand Tour* through the Length and Breadth of British Geography." Area 12: 1980, 245-46.

Davis, William Morris. "An Inductive Study of the Content of Geography." *Bulletin of the American Geographical Society* 38 (1906): 67-84.

Dillon, John B. *A History of Indiana, from Its Earliest Exploration by Europeans to the Close of the Territorial Government, in 1816* (Indianapolis: Bingham & Doughty, 1859).

Fenneman, Nevin M. "Physiographic Boundaries Within the United States." *Annals of the Association of American Geographers* 4 (1914): 84-134.

Fields, S. F. *Where Have the Farm Lands Gone?* (Washington, D.C.: National Agricultural Lands Study, 1979).

Fischel, W. A. "The Urbanization of Agricultural Land: A Review of the National Agricultural Lands Study." *Land Economics* 58 (1982): 236-59.

Hartshorne, Richard. "The Nature of Geography: A Critical Survey of Current Thought in the Light of the Past." *Annals of the Association of American Geographers* 29 (1939): 171-645.

Jordan, Terry G. "Review of *The Look of the Land*." *Annals of the Association of American Geographers* 4 (1975): 579–80.

Kleckner, R. "Trip Report, Iowa State University Statistics Laboratory, September 24, 1980." Unpublished report, Office of Geographic Research, U.S. Geological Survey, Reston, Virginia.

Marschner, F. J. *Land Use and Its Patterns in the United States*. Agricultural Handbook No. 153 (Washington, DC: U.S. Department of Agriculture, 1959).

Medvitz, A. G. "Urban Growth Squeezes California Agriculture," *California Agriculture* 52, 3 (1998): 8-9.

NRCS [Natural Resources Conservation Service] . *Summary Report, 1997*

National Resources Inventory. (Washington, DC: U.S. Department of Agriculture, Natural Resources Conservation Service, 1997; revised December 2000).

Raup, P. M. "An Agricultural Critique of the National Agricultural Lands Study." *Land Economics* 58 (1982): 260-74.

Sauer, Carl O. *The Morphology of Landscape*. Berkeley: University of California Publications in Geography, Vol. 2, No. 2 (1925): 315-50.

Spate, O. H. K, and A. T. A. Learmonth, *India and Pakistan: A General and Regional Geography, 3rd ed.* (London: Methuen, 1967), 407-10.

Stamp, L. Dudley. *The Land of Britain: Its Use and Misuse* (London: Methuen, 1948).

Tavener, L. E. "Changes in the Agricultural Geography of Dorset, 1929-49." *Transactions and Papers*, Institute of British Geographers (1952), 96.

BOOKS BY JOHN FRASER HART

Southeastern Excursion Guidebook, with Eugene Cotton Mather, Publication No. 3 of the XVIIth International Geographical Congress (Washington, DC: International Geographical Union, 1952).

The British Moorlands: A Problem in Land Utilization (Athens: The University of Georgia Press, 1955).

The Southeastern United States, Searchlight Book No. 34 (Princeton: Van Nostrand, 1967).

U.S. and Canada (Boston: Ginn, 1967).

Regions of the United States, editor (New York: Harper and Row, 1972).

The Look of the Land (Englewood Cliffs, NJ: Prentice Hall, 1975).

The South (New York: Van Nostrand, 1976).

Cultural Geography on Topographic Maps, with Karl Raitz (New York: John Wiley, 1975).

The Land that Feeds Us (New York: W. W. Norton, 1991).

Our Changing Cities, editor (Baltimore: The Johns Hopkins University Press, in association with the Center for American Places, 1991).

The Rural Landscape (Baltimore: The Johns Hopkins University Press, in association with the Center for American Places, 1998).

The American Farm, adapted by Gail Kinn (New York: Barnes and Noble, 1998).

The Unknown World of the Mobile Home, with Michelle J. Rhodes and John T. Morgan (Baltimore: The Johns Hopkins University Press, in association with the Center for American Places, 2002).

The Changing Scale of American Agriculture (Charlottesville: University of Virginia Press, 2003).

Landscapes of Minnesota—A Geography, with Susy S. Ziegler (St. Paul: Minnesota Historical Society Press, 2008).

My Kind of County: Door County, Wisconsin (Chicago: The Center for American Places at Columbia College Chicago, 2008).

LIST OF TEXT ILLUSTRATIONS

INDEX

ABOUT THE EDITOR

John C. Hudson is Professor of Geography at Northwestern University in Evanston, Illinois, where he also directs the Environmental Sciences Program. Born in 1941, he is a native of Milton, Wisconsin, and he completed his B.S. in geography the University of Wisconsin-Madison, and, after earning his M.A. and Ph.D. degrees in geography at the University of Iowa, he taught briefly at several universities before joining the Northwestern faculty in 1971. He has been a John Simon Guggenheim Memorial Foundation Fellow, and he served as editor of the *Annals of the Association of American Geographers* from 1976 through 1981. He is the author of *Geographical Diffusion Theory* (1972), *Plains Country Towns* (1985), which received the first John Brinkerhoff Jackson Prize from the Association of American Geographers in 1985, *Crossing the Heartland: Chicago to Denver* (1992), *Making the Corn Belt: A Geographical History of Middle-Western Agriculture* (1994), *Across This Land: A Regional Geography of the United States and Canada* (2002), and *Chicago: A Geography of The City and Region* (2006).

ABOUT THE BOOK:

A *Love of the Land: Selected Writings of John Fraser Har*t was brought to publication as a certified "green book" in an edition of 350 hardcover and 2, 150 paperback copies, with the generous financial support of the Friends of the Center for American Places, for which the publisher is most grateful. All maps, photographs, and other illustrations were made by John Fraser Hart (JFH), except as noted. The text was set in Goudy Old Style. The paper is Chinese Goldeast Matte Book, 120 gsm weight. The book was printed and bound in China.

FOR THE CENTER FOR AMERICAN PLACES AT COLUMBIA COLLEGE CHICAGO:

George F. Thompson, Founder and Director
Brandy L. Savarese, Associate Editorial Director
A. Lenore Lautigar, Associate Editor and Publishing Liason
Sara E. Lovelace and Elizabeth S. Dattilio, Editorial Assistants
Ashleigh M. Frank and Purna Makaram, Manuscript Editors
Laura Ann Leavitt, Indexer
Dawn Hachenski, Book Designer
David Skolkin, Art and Production Director

Center for American Places

AT COLUMBIA COLLEGE CHICAGO

The Center for American Places at Columbia College Chicago is a nonprofit organization, founded in 1990 by George F. Thompson, whose educational mission is to enhance the public's understanding of, appreciation for, and affection for the places of the Americas and the world—whether urban, suburban, rural, or wild. Underpinning this mission is the belief that books provide an indispensable foundation for comprehending and caring for the places where we live, work, and commune. Books live. Books endure. Books make a difference. Books are gifts to civilization.

Since 1990 the Center for American Places at Columbia College Chicago has brought to publication more than 320 books under its own imprint and in association with numerous publishing partners. Center books have won or shared more than 100 editorial awards and citations, including multiple best-books honors in more than thirty fields of study.

For more information, please send inquiries to the Center for American Places at Columbia College Chicago, 600 South Michigan Avenue, Chicago, Illinois, 60605-1996, U.S.A., or visit the Center's Website (www.americanplaces.org).